T0371909

Multi-State Survival Models for Interval-Censored Data

MONOGRAPHS ON STATISTICS AND APPLIED PROBABILITY

General Editors

F. Bunea, V. Isham, N. Keiding, T. Louis, R. L. Smith, and H. Tong

1. Stochastic Population Models in Ecology and Epidemiology *M.S. Barlett* (1960)
2. Queues *D.R. Cox and W.L. Smith* (1961)
3. Monte Carlo Methods *J.M. Hammersley and D.C. Handscomb* (1964)
4. The Statistical Analysis of Series of Events *D.R. Cox and P.A.W. Lewis* (1966)
5. Population Genetics *W.J. Ewens* (1969)
6. Probability, Statistics and Time *M.S. Barlett* (1975)
7. Statistical Inference *S.D. Silvey* (1975)
8. The Analysis of Contingency Tables *B.S. Everitt* (1977)
9. Multivariate Analysis in Behavioural Research *A.E. Maxwell* (1977)
10. Stochastic Abundance Models *S. Engen* (1978)
11. Some Basic Theory for Statistical Inference *E.J.G. Pitman* (1979)
12. Point Processes *D.R. Cox and V. Isham* (1980)
13. Identification of Outliers *D.M. Hawkins* (1980)
14. Optimal Design *S.D. Silvey* (1980)
15. Finite Mixture Distributions *B.S. Everitt and D.J. Hand* (1981)
16. Classification *A.D. Gordon* (1981)
17. Distribution-Free Statistical Methods, 2nd edition *J.S. Maritz* (1995)
18. Residuals and Influence in Regression *R.D. Cook and S. Weisberg* (1982)
19. Applications of Queueing Theory, 2nd edition *G.F. Newell* (1982)
20. Risk Theory, 3rd edition *R.E. Beard, T. Pentikäinen and E. Pesonen* (1984)
21. Analysis of Survival Data *D.R. Cox and D. Oakes* (1984)
22. An Introduction to Latent Variable Models *B.S. Everitt* (1984)
23. Bandit Problems *D.A. Berry and B. Fristedt* (1985)
24. Stochastic Modelling and Control *M.H.A. Davis and R. Vinter* (1985)
25. The Statistical Analysis of Composition Data *J. Aitchison* (1986)
26. Density Estimation for Statistics and Data Analysis *B.W. Silverman* (1986)
27. Regression Analysis with Applications *G.B. Wetherill* (1986)
28. Sequential Methods in Statistics, 3rd edition *G.B. Wetherill and K.D. Glazebrook* (1986)
29. Tensor Methods in Statistics *P. McCullagh* (1987)
30. Transformation and Weighting in Regression *R.J. Carroll and D. Ruppert* (1988)
31. Asymptotic Techniques for Use in Statistics *O.E. Bandorff-Nielsen and D.R. Cox* (1989)
32. Analysis of Binary Data, 2nd edition *D.R. Cox and E.J. Snell* (1989)
33. Analysis of Infectious Disease Data *N.G. Becker* (1989)
34. Design and Analysis of Cross-Over Trials *B. Jones and M.G. Kenward* (1989)
35. Empirical Bayes Methods, 2nd edition *J.S. Maritz and T. Lwin* (1989)
36. Symmetric Multivariate and Related Distributions *K.T. Fang, S. Kotz and K.W. Ng* (1990)
37. Generalized Linear Models, 2nd edition *P. McCullagh and J.A. Nelder* (1989)
38. Cyclic and Computer Generated Designs, 2nd edition *J.A. John and E.R. Williams* (1995)
39. Analog Estimation Methods in Econometrics *C.F. Manski* (1988)
40. Subset Selection in Regression *A.J. Miller* (1990)
41. Analysis of Repeated Measures *M.J. Crowder and D.J. Hand* (1990)
42. Statistical Reasoning with Imprecise Probabilities *P. Walley* (1991)
43. Generalized Additive Models *T.J. Hastie and R.J. Tibshirani* (1990)
44. Inspection Errors for Attributes in Quality Control *N.L. Johnson, S. Kotz and X. Wu* (1991)
45. The Analysis of Contingency Tables, 2nd edition *B.S. Everitt* (1992)
46. The Analysis of Quantal Response Data *B.J.T. Morgan* (1992)
47. Longitudinal Data with Serial Correlation—A State-Space Approach *R.H. Jones* (1993)

48. Differential Geometry and Statistics *M.K. Murray and J.W. Rice* (1993)
49. Markov Models and Optimization *M.H.A. Davis* (1993)
50. Networks and Chaos—Statistical and Probabilistic Aspects
O.E. Barndorff-Nielsen, J.L. Jensen and W.S. Kendall (1993)
51. Number-Theoretic Methods in Statistics *K.-T. Fang and Y. Wang* (1994)
52. Inference and Asymptotics *O.E. Barndorff-Nielsen and D.R. Cox* (1994)
53. Practical Risk Theory for Actuaries *C.D. Daykin, T. Pentikäinen and M. Pesonen* (1994)
54. Biplots *J.C. Gower and D.J. Hand* (1996)
55. Predictive Inference—An Introduction *S. Geisser* (1993)
56. Model-Free Curve Estimation *M.E. Tarter and M.D. Lock* (1993)
57. An Introduction to the Bootstrap *B. Efron and R.J. Tibshirani* (1993)
58. Nonparametric Regression and Generalized Linear Models *P.J. Green and B.W. Silverman* (1994)
59. Multidimensional Scaling *T.F. Cox and M.A.A. Cox* (1994)
60. Kernel Smoothing *M.P. Wand and M.C. Jones* (1995)
61. Statistics for Long Memory Processes *J. Beran* (1995)
62. Nonlinear Models for Repeated Measurement Data *M. Davidian and D.M. Giltinan* (1995)
63. Measurement Error in Nonlinear Models *R.J. Carroll, D. Rupert and L.A. Stefanski* (1995)
64. Analyzing and Modeling Rank Data *J.J. Marden* (1995)
65. Time Series Models—In Econometrics, Finance and Other Fields
D.R. Cox, D.V. Hinkley and O.E. Barndorff-Nielsen (1996)
66. Local Polynomial Modeling and its Applications *J. Fan and I. Gijbels* (1996)
67. Multivariate Dependencies—Models, Analysis and Interpretation *D.R. Cox and N. Wermuth* (1996)
68. Statistical Inference—Based on the Likelihood *A. Azzalini* (1996)
69. Bayes and Empirical Bayes Methods for Data Analysis *B.P. Carlin and T.A Louis* (1996)
70. Hidden Markov and Other Models for Discrete-Valued Time Series *I.L. MacDonald and W. Zucchini* (1997)
71. Statistical Evidence—A Likelihood Paradigm *R. Royall* (1997)
72. Analysis of Incomplete Multivariate Data *J.L. Schafer* (1997)
73. Multivariate Models and Dependence Concepts *H. Joe* (1997)
74. Theory of Sample Surveys *M.E. Thompson* (1997)
75. Retrial Queues *G. Falin and J.G.C. Templeton* (1997)
76. Theory of Dispersion Models *B. Jørgensen* (1997)
77. Mixed Poisson Processes *J. Grandell* (1997)
78. Variance Components Estimation—Mixed Models, Methodologies and Applications *P.S.R.S. Rao* (1997)
79. Bayesian Methods for Finite Population Sampling *G. Meeden and M. Ghosh* (1997)
80. Stochastic Geometry—Likelihood and computation
O.E. Barndorff-Nielsen, W.S. Kendall and M.N.M. van Lieshout (1998)
81. Computer-Assisted Analysis of Mixtures and Applications—Meta-Analysis, Disease Mapping and Others
D. Böhning (1999)
82. Classification, 2nd edition *A.D. Gordon* (1999)
83. Semimartingales and their Statistical Inference *B.L.S. Prakasa Rao* (1999)
84. Statistical Aspects of BSE and vCJD—Models for Epidemics *C.A. Donnelly and N.M. Ferguson* (1999)
85. Set-Indexed Martingales *G. Ivanoff and E. Merzbach* (2000)
86. The Theory of the Design of Experiments *D.R. Cox and N. Reid* (2000)
87. Complex Stochastic Systems *O.E. Barndorff-Nielsen, D.R. Cox and C. Klüppelberg* (2001)
88. Multidimensional Scaling, 2nd edition *T.F. Cox and M.A.A. Cox* (2001)
89. Algebraic Statistics—Computational Commutative Algebra in Statistics
G. Pistone, E. Riccomagno and H.P. Wynn (2001)
90. Analysis of Time Series Structure—SSA and Related Techniques
N. Golyandina, V. Nekrutkin and A.A. Zhigljavsky (2001)
91. Subjective Probability Models for Lifetimes *Fabio Spizzichino* (2001)
92. Empirical Likelihood *Art B. Owen* (2001)
93. Statistics in the 21st Century *Adrian E. Raftery, Martin A. Tanner, and Martin T. Wells* (2001)
94. Accelerated Life Models: Modeling and Statistical Analysis
Vilijandas Bagdonavicius and Mikhail Nikulin (2001)

95. Subset Selection in Regression, Second Edition *Alan Miller* (2002)
96. Topics in Modelling of Clustered Data *Marc Aerts, Helena Geys, Geert Molenberghs, and Louise M. Ryan* (2002)
97. Components of Variance *D.R. Cox and P.J. Solomon* (2002)
98. Design and Analysis of Cross-Over Trials, 2nd Edition *Byron Jones and Michael G. Kenward* (2003)
99. Extreme Values in Finance, Telecommunications, and the Environment *Bärbel Finkenstädt and Holger Rootzén* (2003)
100. Statistical Inference and Simulation for Spatial Point Processes *Jesper Møller and Rasmus Plenge Waagepetersen* (2004)
101. Hierarchical Modeling and Analysis for Spatial Data *Sudipto Banerjee, Bradley P. Carlin, and Alan E. Gelfand* (2004)
102. Diagnostic Checks in Time Series *Wai Keung Li* (2004)
103. Stereology for Statisticians *Adrian Baddeley and Eva B. Vedel Jensen* (2004)
104. Gaussian Markov Random Fields: Theory and Applications *Håvard Rue and Leonhard Held* (2005)
105. Measurement Error in Nonlinear Models: A Modern Perspective, Second Edition *Raymond J. Carroll, David Ruppert, Leonard A. Stefanski, and Ciprian M. Crainiceanu* (2006)
106. Generalized Linear Models with Random Effects: Unified Analysis via H-likelihood *Youngjo Lee, John A. Nelder, and Yudi Pawitan* (2006)
107. Statistical Methods for Spatio-Temporal Systems *Bärbel Finkenstädt, Leonhard Held, and Valerie Isham* (2007)
108. Nonlinear Time Series: Semiparametric and Nonparametric Methods *Jiti Gao* (2007)
109. Missing Data in Longitudinal Studies: Strategies for Bayesian Modeling and Sensitivity Analysis *Michael J. Daniels and Joseph W. Hogan* (2008)
110. Hidden Markov Models for Time Series: An Introduction Using R *Walter Zucchini and Iain L. MacDonald* (2009)
111. ROC Curves for Continuous Data *Wojtek J. Krzanowski and David J. Hand* (2009)
112. Antedependence Models for Longitudinal Data *Dale L. Zimmerman and Vicente A. Núñez-Antón* (2009)
113. Mixed Effects Models for Complex Data *Lang Wu* (2010)
114. Intoduction to Time Series Modeling *Genshiro Kitagawa* (2010)
115. Expansions and Asymptotics for Statistics *Christopher G. Small* (2010)
116. Statistical Inference: An Integrated Bayesian/Likelihood Approach *Murray Aitkin* (2010)
117. Circular and Linear Regression: Fitting Circles and Lines by Least Squares *Nikolai Chernov* (2010)
118. Simultaneous Inference in Regression *Wei Liu* (2010)
119. Robust Nonparametric Statistical Methods, Second Edition *Thomas P. Hettmansperger and Joseph W. McKean* (2011)
120. Statistical Inference: The Minimum Distance Approach *Ayanendranath Basu, Hiroyuki Shioya, and Chanseok Park* (2011)
121. Smoothing Splines: Methods and Applications *Yuedong Wang* (2011)
122. Extreme Value Methods with Applications to Finance *Serguei Y. Novak* (2012)
123. Dynamic Prediction in Clinical Survival Analysis *Hans C. van Houwelingen and Hein Putter* (2012)
124. Statistical Methods for Stochastic Differential Equations *Mathieu Kessler, Alexander Lindner, and Michael Sørensen* (2012)
125. Maximum Likelihood Estimation for Sample Surveys *R. L. Chambers, D. G. Steel, Suojin Wang, and A. H. Welsh* (2012)
126. Mean Field Simulation for Monte Carlo Integration *Pierre Del Moral* (2013)
127. Analysis of Variance for Functional Data *Jin-Ting Zhang* (2013)
128. Statistical Analysis of Spatial and Spatio-Temporal Point Patterns, Third Edition *Peter J. Diggle* (2013)
129. Constrained Principal Component Analysis and Related Techniques *Yoshio Takane* (2014)
130. Randomised Response-Adaptive Designs in Clinical Trials *Anthony C. Atkinson and Atanu Biswas* (2014)
131. Theory of Factorial Design: Single- and Multi-Stratum Experiments *Ching-Shui Cheng* (2014)
132. Quasi-Least Squares Regression *Justine Shults and Joseph M. Hilbe* (2014)
133. Data Analysis and Approximate Models: Model Choice, Location-Scale, Analysis of Variance, Nonparametric Regression and Image Analysis *Laurie Davies* (2014)
134. Dependence Modeling with Copulas *Harry Joe* (2014)
135. Hierarchical Modeling and Analysis for Spatial Data, Second Edition *Sudipto Banerjee, Bradley P. Carlin, and Alan E. Gelfand* (2014)

136. Sequential Analysis: Hypothesis Testing and Changepoint Detection *Alexander Tartakovsky, Igor Nikiforov, and Michèle Basseville* (2015)
137. Robust Cluster Analysis and Variable Selection *Gunter Ritter* (2015)
138. Design and Analysis of Cross-Over Trials, Third Edition *Byron Jones and Michael G. Kenward* (2015)
139. Introduction to High-Dimensional Statistics *Christophe Giraud* (2015)
140. Pareto Distributions: Second Edition *Barry C. Arnold* (2015)
141. Bayesian Inference for Partially Identified Models: Exploring the Limits of Limited Data *Paul Gustafson* (2015)
142. Models for Dependent Time Series *Granville Tunnicliffe Wilson, Marco Reale, John Haywood* (2015)
143. Statistical Learning with Sparsity: The Lasso and Generalizations *Trevor Hastie, Robert Tibshirani, and Martin Wainwright* (2015)
144. Measuring Statistical Evidence Using Relative Belief *Michael Evans* (2015)
145. Stochastic Analysis for Gaussian Random Processes and Fields: With Applications *Vidyadhar S. Mandrekar and Leszek Gawarecki* (2015)
146. Semialgebraic Statistics and Latent Tree Models *Piotr Zwiernik* (2015)
147. Inferential Models: Reasoning with Uncertainty *Ryan Martin and Chuanhai Liu* (2016)
148. Perfect Simulation *Mark L. Huber* (2016)
149. State-Space Methods for Time Series Analysis: Theory, Applications and Software *Jose Casals, Alfredo Garcia-Hiernaux, Miguel Jerez, Sonia Sotoca, and A. Alexandre Trindade* (2016)
150. Hidden Markov Models for Time Series: An Introduction Using R, Second Edition *Walter Zucchini, Iain L. MacDonald, and Roland Langrock* (2016)
151. Joint Modeling of Longitudinal and Time-to-Event Data *Robert M. Elashoff, Gang Li, and Ning Li* (2016)
152. Multi-State Survival Models for Interval-Censored Data *Ardo van den Hout* (2016)

Monographs on Statistics and Applied Probability 152

Multi-State Survival Models for Interval-Censored Data

Ardo van den Hout

Department of Statistical Science
University College London, UK

CRC Press
Taylor & Francis Group
Boca Raton London New York

CRC Press is an imprint of the
Taylor & Francis Group, an **informa** business
A CHAPMAN & HALL BOOK

CRC Press
Taylor & Francis Group
6000 Broken Sound Parkway NW, Suite 300
Boca Raton, FL 33487-2742

Printed on acid-free paper
Version Date: 20160725

International Standard Book Number-13: 978-1-4665-6840-2 (Hardback)

Visit the Taylor & Francis Web site at
http://www.taylorandfrancis.com

and the CRC Press Web site at
http://www.crcpress.com

For Marije

Contents

Preface

This book is about statistical inference using multi-state survival models. The aim is to introduce and explain methods to describe stochastic processes that consist of transitions between states over time. The book is targeted at applications in medical statistics, epidemiology, demography, and social statistics. An example of an application is a three-state process for dementia and survival in the older population. Such a process can be described by an illness-death model, where state 1 is the dementia-free state, state 2 is the dementia state, and state 3 is the dead state. Statistical analysis can investigate potential associations between the risk of moving to the next state and variables such as age, gender, or education. Statistical analysis can also be used to predict the multi-state survival process. A prediction of specific interest is residual life expectancy; that is, prediction of the expected number of years of life remaining at a given age. When the model describes an illness-death process in the older population, total residual life expectancy at a given age can be subdivided into healthy life expectancy and life expectancy in ill-health.

Applications in this book concern longitudinal data. Typically the data are subject to interval censoring in the sense that some of the transition times are not observed but are known to lie within a given time interval. For example, the time of onset of dementia is latent but when longitudinal data are available the onset may be known to lie in the time interval defined by two successive observations.

Methodologically, multi-state modelling is an elegant combination of statistical inference and the theory of stochastic processes. With this book, I aim to show that the statistical modelling is versatile and allows for a wide range of applications. The computation that is involved in fitting the models can be considerable, but in many cases existing software can be utilised.

I hope that the book will be of interest to diverse groups of readers. Firstly, the book introduces multi-state survival modelling for researchers who are new to the subject. After the introductory first chapter, the book discusses the multi-state survival model as an extension of the (two-state) standard survival model. Further topics are discrete-time models versus continuous-time models, theory on continuous-time Markov chains, parametric models for transition-specific baseline hazards, maximum likelihood inference, model validation, and prediction. The appendix includes code for the R software.

Secondly, for readers with subject-matter knowledge, the book can serve as a reference and it also offers advanced topics such as the modelling of time-dependent hazards, a general scoring algorithm for maximum likelihood inference, methods for Bayesian inference, estimation of state-specific life expectancies, semi-parametric models, and specification and estimation of frailty models.

For the software used in the data analysis, some details are provided in Appendix C. Software is also available on the author's website, `www.ucl.ac.uk/~ucakadl/Book`.

The book assumes knowledge of mathematical statistics at third-year undergraduate or at MSc level. This knowledge can be based on courses in statistics or applied mathematics, but also on courses in other disciplines with a strong statistical programme.

Acknowledgments

Thinking about the statistical methods, writing about them, coding the computation, and analysing data have been very much a joint effort in the past years. Discussions with collaborators have helped me to understand both the scope and the finer details of multi-state survival models. Special thanks to Fiona Matthews for supervising my first years of research on this topic, and for all the joint work. I would also like to thank my collaborators Tirza Buter, Jean-Paul Fox, Jutta Gampe, Carol Jagger, Venediktos Kapetanakis, Rinke Klein Entink, Robson Machado, Riccardo Marioni, Graciela Muniz-Terrera, Ekaterina Ogurtsova, Nora Pashayan, Luz Sanchez-Romero, and Brian Tom.

I am grateful to the Medical Research Council Biostatistics Unit in Cambridge, and the Department of Statistical Science, University College London, for allowing me time for research. I would like to thank the colleagues at these two institutes who helped my research with discussions and feedback. Thanks also to those students at University College London who did research projects on multi-state models. These projects have been an additional source for learning and insight.

I am in debt to the people who made it possible to use the longitudinal data for the examples in this book. This includes both the participants in the studies and the teams who collected and managed the data; see Section 1.5.

Thanks to Rob Calver at CRC Press for encouragement to write this book, and for his help and patience. Finally, many thanks to the two anonymous reviewers of the book manuscript. Their detailed and substantial remarks have been a great help in improving the material.

Chapter 1

Introduction

1.1 Multi-state survival models

Multi-state models are routinely used in research where change of status over time is of interest. In epidemiology and medical statistics, for example, the models are used to describe health-related processes over time, where status is defined by a disease or a condition. In social statistics and in demography, the models are used to study processes such as region of residence, work history, or marital status. A multi-state model that includes a dead state is called a *multi-state survival model.*

The specification and estimation of a multi-state survival model depend partly on the study design which generated the longitudinal data that are under investigation. An important distinction is whether or not exact times are observed for transitions between the states. This book considers mainly study designs where death times are known (or right-censored) and where times of transitions between the living states are interval-censored. Many applications in epidemiology and medical statistics have this property as it is often hard to measure the exact time of onset of a disease or condition. Examples are dementia, cognitive decline, disability in old age, and infectious diseases.

Closely related to the interval censoring is the choice between discrete-time and continuous-time models. A discrete-time model assumes a stepwise transition process, where the fixed time between successive steps is not part of the model. There are applications where this model is appropriate, and other applications for which this model is a good approximation of the process of interest. Continuous-time models, the topic of this book, allow changes of state at any time and will be more realistic in many situations. This type of model is also more flexible with respect to the study design for the observation times.

A continuous-time multi-state model can be seen as an extension of the standard survival model. The latter can be defined as a two-state model where a one-off change of status is the event of interest. The archetype example in medical statistics is the transition from the state of being alive to the state of being dead. Often there will be additional information in the data on other stochastic events that may be associated with the risk of a transition. An example of this is the onset of dementia in a study of survival in the older population. In such a case, the onset of dementia can be taken into account in the survival model by including it as binary time-dependent covariate for the risk of dying. The multi-state model approach would in this case consist of defining three states: a dementia-free state, a dementia state, and a dead state.

Assuming that both models are parametric, there are some clear advantages of the three-state model over the two-state model. Firstly, because the onset of dementia is part of the model, the three-state model can be used for prediction. Even though the two-state model can be used to study the effect of dementia on survival, this model cannot be used for prediction without additional modelling of the stochastic process which underlies the onset of dementia. Secondly, the onset of dementia is a latent process—observation will always be interval-censored. Dealing with this interval censoring is not straightforward when the onset is included in the two-state model as a time-dependent covariate. The multi-state models in this book, however, are explicitly defined for interval-censored transitions between living states.

Even in applications where survival is not of immediate interest, multi-state survival model can still be very useful. Especially in epidemiological and medical research, a longitudinal outcome variable of interest may be correlated with survival. If this is the case, the risk of dying cannot be ignored in the data analysis. Examples of such longitudinal outcomes are blood pressure, cognitive function in the older population, and biomarkers for cardiovascular disease. One option is to specify a model for the longitudinal process of interest and combine this model with a standard survival model. This is called a joint model, and it is often specified with random effects which are shared by the two constituent models. But if the longitudinal outcome can be adequately discretised by a set of living states, then a multi-state survival model can be defined by adding a dead state. This provides an elegant alternative to a joint model: the multi-state model can be defined as an overall fixed-effects model for the process of interest as well as for survival.

A multi-state process is called a *Markov chain* if all information about the future is contained in the present state. If, for example, time spent in the present state affects the risk of a transition to the next state, then the process is not a Markov chain. Although many processes in real life are not Markovian, a statistical model based upon a Markov chain may still provide a good approximation of the process of interest. Most of the multi-state models in this book are not Markovian in the strict sense. By linking age and values of covariates to the risk of a transition, information about the future is contained not only in the present state, but also in current age and additional background characteristics.

This first chapter introduces the basic concepts that are used in multi-state survival models, discusses the relevant type of data, and illustrates the scope of the statistical modelling by an example. Details of parameter estimation and statistical inference are postponed to later chapters. At the end of this chapter an overview will be given of the methods and the examples in the rest of the book.

1.2 Basic concepts

The standard two-state survival model is defined by distinguishing a living state and dead state. The two main features of the standard survival model are that there is one event of interest (the transition from alive to dead) and that the timing of this event may be right-censored, in which case it is known that the event has not happened yet.

As an example, say patients are followed up for a year after a risky medical operation and the time scale is months since the operation. When the event is defined by death, the information per patient is either time of death or the time at which death is right-censored. The latter is typically the end of the follow-up, in this example 12 months. Alternatively, the patient drops out of the study during the follow-up and the censored time is the last time the patient was seen alive. The statistical modelling of survival in the presence of right censoring is often undertaken by assuming a model for the hazard. The hazard is the instantaneous risk of the event. As a quantity it is an unbounded positive value and should be distinguished from a probability, which is a value between zero and one. In the example, the probability of the event is linked to a specified time interval, for example, the probability of dying within a year, whereas the hazard is defined for a moment in time specified in months.

In a multi-state survival model there is more than one event of interest. An example with two states is the functioning of a machine with working (state 1) versus being repaired (state 2). The events are the transition from state 1 to state 2 and the transition from state 2 to state 1. An example with three states is an illness-death model where state 1 is the dementia-free state, state 2 is the state with dementia, and state 3 is the dead state. This defines three events: death from state 1, death from state 2, and onset of dementia.

A multi-state survival process can be described by a model for the transition-specific hazards. There are as many hazards to model as there are transitions. Censoring is often more complex than in the standard survival model. In the example with the dementia state, for an individual who is last seen in state 1, right censoring at a later time concerns both death and a potential transition to the dementia state. Furthermore, often it is known that a transition between states took place without knowing the exact time of the transition. This is called *interval censoring.* The onset of dementia between two screening times is an example of this.

For the formal definition of interval censoring, the following summarises the discussion in Sun (2006). Let random variable T denote the time to the event. The sample space of T is $[0,\infty)$. An observation on T is interval-censored if only an interval $(L,R]$ is observed such that

$$T \in (L,R].$$

This notation extends to exact observations of T in case $L = R$, and to right-censored observations in case $R = \infty$. Interval-censored data in this book refer to the situation where the data include at least one interval $(L,R]$ with $L \neq R$ and both L and R in $(0,\infty)$. In the case of longitudinal data where multiple events are of interest, interval-censored data are also called *panel data* (Kalbfleisch and Lawless, 1985), or *intermittently observed data* (Tom and Farewell, 2011; Cook and Lawless, 2014).

Independent interval censoring implies that the mechanism that generates the censoring is independent of T. An example of this for longitudinal data is when all observation times during follow-up are determined in advance at baseline. Formally, the interval censoring is independent when

$$P(L < T \leq R | L = l, R = r) = P(l < T \leq r),$$

that is, when the joint distribution of L and R is free of the parameters involved in the distribution of T (Sun, 2006, Section 1.3). If the interval censoring is independent, then the censoring mechanism can be ignored in the data analysis.

In biostatistics, complete independence between observation times and a multi-state disease process is often not realistic. Given random variable M for the number of observations, Grüger et al. (1991) define an observation scheme $\{T_1,\ldots,T_m | M = m\}$ to be *non-informative* for the multi-state disease process if the full likelihood on $\{T_1,\ldots,T_m | M = m\}$ is proportional to the likelihood obtained if the number of observations and their times were fixed in advance. It follows that if the observation scheme is non-informative, then the scheme can be ignored when maximum likelihood is used to estimate parameters for the multi-state process.

Independent interval censoring implies a non-informative observation scheme but not vice versa. An important and illustrative example of an observation scheme that induces dependent interval censoring but is non-informative is *Doctor's care* as defined in Grüger et al. (1991). In this scheme, a next observation time is chosen on the basis of the current state. It states are defined with respect to disease progression, this would imply shorter time intervals between successive observations for patients with increased risk of dying, and longer intervals for patients in the healthy state or in a stable stage of the disease.

An example of an observation scheme that is not non-informative is patient self-selection. In this scenario, patients go to the doctor because of their disease progression. If such a visit is included as an observation of the state in the longitudinal data, then this leads to biased likelihood inference when the self-selection is not taken into account.

1.3 Example

1.3.1 Cardiac allograft vasculopathy (CAV) study

To illustrate the statistical modelling of a multi-state survival process, we use data with individual histories of angiographic examinations of 622 heart transplant recipients. Data were collected at Papworth Hospital in the United Kingdom and are included in the R package msm (Jackson, 2011). The data are available for illustration purposes only and should not be used in isolation to inform clinical practice. Permission to use the data for this book was kindly given by Steven Tsui at Papworth Hospital.

Cardiac allograft vasculopathy (CAV) is a deterioration of the arterial walls. For the heart transplant recipients, four states are defined: no CAV, mild/moderate CAV, severe CAV, and dead; see Sharples et al. (2003). Figure 1.1 depicts the states and transitions in the data.

Yearly examinations for CAV took place after the transplant for up to 18 years. There is variation in time between examinations: it was not always exactly 1 year and some patients skipped one or more scheduled examinations. For example, the patient who was followed up for 18 years and whose death is right-censored has observed states at 0, 1.01, 2.01, 3.01, 4.07, 5.00, 6.01, 7.92, 8.92, 11.03, 12.01, 15.63, and 17.98 years after transplant. For the whole sample, the follow-up in years until right censoring or death is given in Figure 1.2. The histogram shows that a follow-up with observation during the whole period of 18 years is exceptional. The mean of the follow-up time is 5.9 years and the median is 5.0 years.

Transitions between the living states are interval-censored, but if the patient died during follow-up, then his or her death time is known exactly.

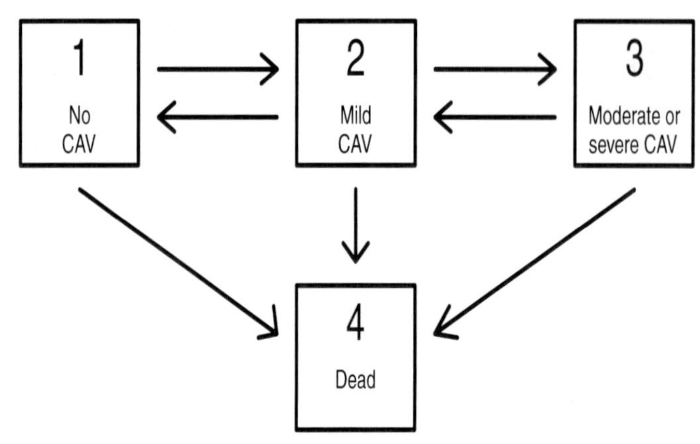

Figure 1.1 *Transitions in the data for the four-state model for cardiac allograft vasculopathy (CAV).*

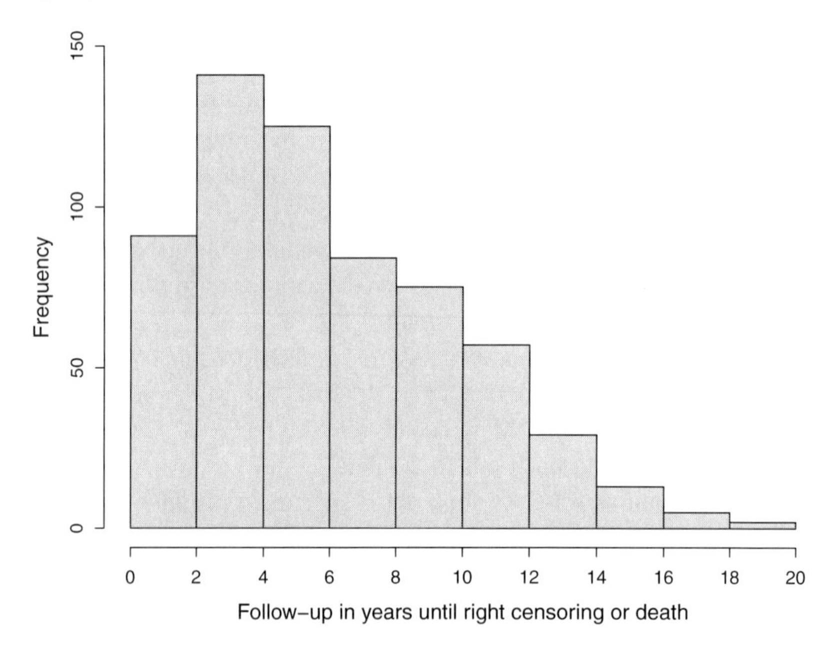

Figure 1.2 *For the CAV data, frequencies of follow-up time in years until right censoring or death.*

Figure 1.3 shows how observed state changes over time. The sizes of the diameters correspond with the observed frequency in the corresponding state. Due to the interval censoring, the information in Figure 1.3 is limited. For example, the state of the individual who was followed up for 18 years

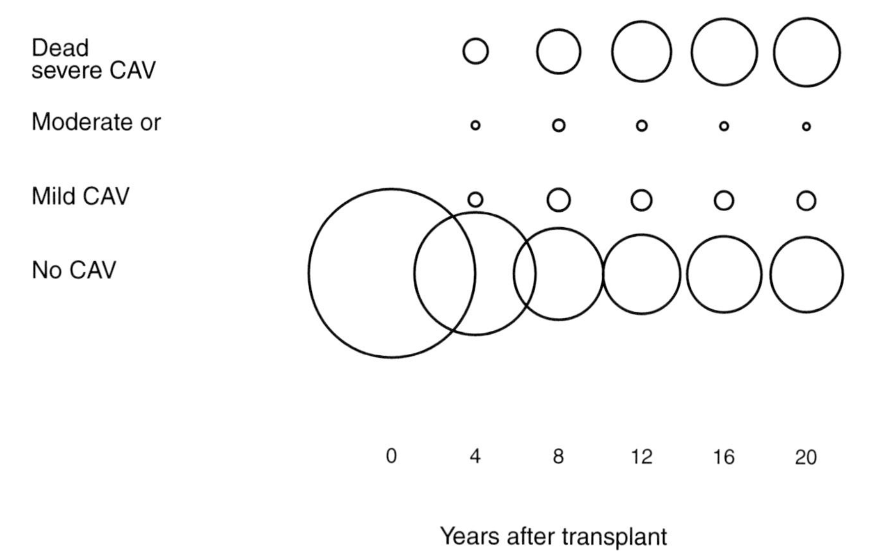

Figure 1.3 *For the CAV data, last seen state for an imposed time grid in years after transplant. Difference in diameters refers to difference in frequencies. All* $N = 622$ *individuals start in the state without CAV.*

and whose observation times are given above was not observed at year 16. At 15.63 years after transplant, he or she did not have CAV, but at 17.98 years mild CAV was observed. Whether this individual obtained CAV before 16 years after transplant or later is not known. When we do not have exact transition times, we only know the state that was observed at the last time an individual was seen. Hence, the diagram is only an approximation of the true status of the process at the specified time points.

Despite the fact that the diagram in Figure 1.3 is an approximation it gives a good idea of how the process develops over time. It shows that all individuals start without CAV and that most of the movement between the states takes place in the first 12 years. Note that the diagram cannot distinguish whether the states with CAV are relatively rare or that the duration in these states is short. The dead state is absorbing and the frequency for that state does not decrease. The frequency for the no-CAV state does not increase over time, which agrees with the general idea of CAV being progressive.

Further information about the CAV process is given by the frequencies in Table 1.1. This is called the *state table* and is a way to summarise multi-state data. The frequencies are the number of times each pair of states was observed at successive observation times. For the CAV data, the diagonal for the living states dominates which illustrates that the process of change is slow relative

Table 1.1 *State table for the CAV data: number of times each pair of states was observed at successive observation times.*

	To			
From	No CAV	Mild CAV	Mod./Severe CAV	Dead
No CAV	1367	204	44	148
Mild CAV	46	134	54	48
Mod./severe CAV	4	13	107	55

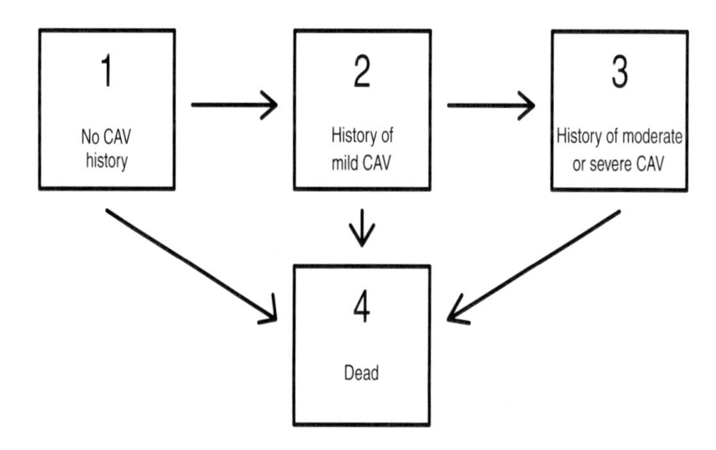

Figure 1.4 *Four-state progressive model for history of cardiac allograft vasculopathy (CAV).*

to the timing of the follow-up. Note that from the moderate/severe state there are just a few observed backward transitions, which shows the progressive nature of the deterioration of the arterial walls.

1.3.2 A four-state progressive model

Sharples et al. (2003) assume that CAV is a progressive process where recovery is not possible and that observed recoveries are due to misclassification of states. Multi-state models have been developed for this situation. Sharples et al. (2003) use such a model to analyse the CAV data by distinguishing a latent progressive multi-state process from the observed manifest process. More information on how misclassification can be taken into account in multi-state survival models is given in Section 8.5, which includes an example with the CAV data.

For the current analysis, we define a progressive multi-state model for observed history of CAV. The time scale t in the model is *time in years since transplant*. CAV history at time t is defined as the highest CAV state observed up to and including t. This implies that there are five transitions; see Figure 1.4. States 1, 2, and 3 are defined as the states without CAV history, with a history of mild CAV only, and with a history of moderate/severe CAV, respectively. State 4 is the dead state.

For each of the five transitions the hazard is modelled using log-linear regression equations. In these equations, the covariates are time since transplant (t), baseline age at transplant ($b.age$), and age of the donor ($d.age$). The hazard for the transition from state r to state s at time t is denoted by $q_{rs}(t)$, and the regression equations are given by

$$\begin{aligned}
q_{12}(t) &= \exp(\beta_{12.0} + \beta_{12.1}t + \beta_{12.2}\ b.age + \beta_{12.3}\ d.age) \\
q_{14}(t) &= \exp(\beta_{14.0} + \beta_{14.1}t + \beta_{14.2}\ b.age + \beta_{14.3}\ d.age) \\
q_{23}(t) &= \exp(\beta_{23.0} + \beta_{23.3}\ d.age) \\
q_{24}(t) &= \exp(\beta_{24.0} + \beta_{24.3}\ d.age) \\
q_{34}(t) &= \exp(\beta_{34.0} + \beta_{34.3}\ d.age).
\end{aligned} \tag{1.1}$$

Many more models can be defined for the CAV-history data. Specification (1.1) is used as an example. It reflects our interest in the age of the donor as a risk factor for leaving state 1. To control for the effects of time since transplant and baseline age, these two covariates are also included in the relevant regression equations.

Maximum likelihood can be used to estimate the model parameters. The details for defining the likelihood function are presented in Section 4.5. An important feature of the statistical inference for (1.1) is that piecewise-constant hazards are used in the maximum likelihood estimation. The piecewise-constant hazards are an approximation of the continuous-time hazard specification in (1.1).

Grid points for the approximation can be defined by the data; that is, the grid in the likelihood contribution for a given individual is constructed using individual observation times. Given the survey design of yearly observations, this implies in most cases that the approximation of the hazard allows a stepwise change from year to year. If an individual skips one planned observation, the approximation is cruder and the hazard stays constant for 2 years. For example, the likelihood contribution for the individual who was mentioned above and who was followed up for 18 years is defined using constant hazards within the intervals $(0, 1.01]$, $(1.01, 2.01]$, $(2.01, 3.01]$, etc.

Table 1.2 *Parameter estimates for the four-state model for the CAV-history data. Estimated standard errors in parentheses. Intercepts, and effects of time since transplant (t), age at transplant (b.age), and age of the donor (d.age).*

Intercept		t		d.age	
$\beta_{12.0}$	−3.476 (0.356)	$\beta_{12.1}$	0.118 (0.025)	$\beta_{12.3}$	0.026 (0.006)
$\beta_{14.0}$	−6.427 (0.724)	$\beta_{14.1}$	0.019 (0.051)	$\beta_{14.3}$	0.023 (0.010)
$\beta_{23.0}$	−1.250 (0.297)			$\beta_{23.3}$	−0.006 (0.009)
$\beta_{24.0}$	−1.896 (1.194)	b.age		$\beta_{24.3}$	−0.042 (0.048)
$\beta_{34.0}$	−1.057 (0.388)	$\beta_{12.2}$	0.002 (0.007)	$\beta_{34.3}$	−0.007 (0.013)
		$\beta_{14.2}$	0.052 (0.012)		

Using the piecewise-constant approximation with a grid defined by the data implies that the estimation of the model parameters becomes data-dependent. The model itself still assumes a continuous-time dependency. This can be distinguished from a piecewise-constant hazard model which explicitly assumes that the hazard is constant from one interval to the next.

The main advantage of the piecewise-constant hazard approximation is that it simplifies the estimation of the model parameters and—in this case—that the maximum likelihood estimation can be undertaken using the R package msm (Jackson, 2011). Parameters can also be estimated using the scoring algorithm for multi-state survival models presented in Chapter 4, or using the sampling methods for Bayesian inference in Chapter 6.

Table 1.2 presents the model parameters in (1.1) as estimated by maximum likelihood where the piecewise-constant approximation is defined by the data. An absolute value for the hazard is of limited information on its own, but values can be compared. Table 1.2 shows that with respect to covariate effects, an increase of years since transplant is associated with a higher risk of moving from the state without CAV history to the state with a history of moderate/severe CAV. Regarding the age of the donor, the estimate of the regression coefficient $\beta_{12.3}$ in Table 1.2 shows that a higher donor age is associated with an increase risk of moving out of the state without CAV history. For the states with CAV history, the age of the donor does not seem to have a significant effect.

More informative than hazards are transition probabilities, which are the probabilities of transition between states within a specified time interval. It is common to present these probabilities in a matrix. In the current application, this is a 4×4 matrix, where the rows are the current state and the columns the next state. Assume we are interested in the first year after the transplant.

At the time of the transplant, median age of the patients in the sample is 49.5 years. Median donor age is 28.5. Conditional on these median values, we obtain the probability matrix

$$\widehat{\mathbf{P}}(t) = \begin{pmatrix} \widehat{p}_{11}(t) & \widehat{p}_{12}(t) & \widehat{p}_{13}(t) & \widehat{p}_{14}(t) \\ \widehat{p}_{21}(t) & \widehat{p}_{22}(t) & \widehat{p}_{23}(t) & \widehat{p}_{24}(t) \\ \widehat{p}_{31}(t) & \widehat{p}_{32}(t) & \widehat{p}_{33}(t) & \widehat{p}_{34}(t) \\ \widehat{p}_{41}(t) & \widehat{p}_{42}(t) & \widehat{p}_{43}(t) & \widehat{p}_{44}(t) \end{pmatrix} = \begin{pmatrix} 0.894 & 0.060 & 0.007 & 0.039 \\ 0 & 0.752 & 0.181 & 0.067 \\ 0 & 0 & 0.755 & 0.245 \\ 0 & 0 & 0 & 1 \end{pmatrix},$$

where $t = 1$ year. For the first three elements of the diagonal, the 95% confidence intervals are $(0.877, 0.908)$, $(0.707, 0.783)$, and $(0.707, 0.800)$, respectively. For example, the entry $\widehat{p}_{12}(t)$ implies that for someone who is 49.5 years old at the time of transplant and who gets a donor heart from a 28.5 year old has an estimated probability of 0.060 of having a history of mild CAV after 1 year.

As an alternative to the above piecewise-constant approximation where individual data determine the grid for the approximation in the likelihood contribution per individual, a fixed grid can be used that is imposed for all likelihood contributions. This is still an approximation of the time-continuous change of the hazard in (1.1), but there are two advantages. Firstly, the grid is the same across individuals and independent of the data. Secondly, if there are individuals with long time intervals between two successive observations, the imposed grid subdivides those long intervals into shorter ones resulting in a better approximation of the continuous-time dependency.

The choice between using individual grids or an imposed grid can make a difference in the estimation of model parameters. This will depend on the study design in combination with the volatility of the process of interest.

For the CAV data, we fitted the model defined by (1.1) using an imposed grid with time points 0, 2, 4, 6, 10, and 15 years after baseline. Results were similar compared to the previous analysis. For example, the estimated transition matrix for the first year conditional on median age at the time of transplant and median donor age is

$$\widehat{\mathbf{P}}(t) = \begin{pmatrix} 0.896 & 0.059 & 0.007 & 0.039 \\ 0 & 0.750 & 0.181 & 0.069 \\ 0 & 0 & 0.759 & 0.241 \\ 0 & 0 & 0 & 1 \end{pmatrix}.$$

There are alternatives to using a piecewise-constant approximation for estimation of time-dependent models. Chapter 3 discusses using numerical integration for progressive three-state models. Yet another option is presented by Titman (2011) who uses direct numerical solutions to the Kolmogorov Forward equations which are the differential equations that define the multi-state transition probabilities.

Further understanding of the multi-state process can be derived from estimating state-specific length of stay. Given that the model contains a dead state, length of stay can be formulated at residual state-specific life expectancy. In the CAV context, it is of interest how remaining total life expectancy subdivides into life expectancies in the living states 1, 2, and 3, and whether there is heterogeneity depending on covariate values at the time of transplant. Life expectancies can be derived from a fitted model using transition probabilities. This will be discussed in Chapter 7.

The hazard specification in (1.1) is just one of the many ways a continuous-time hazard model can be defined. If we write the first equation in (1.1) as

$$q_{12}(t) = \lambda_{12} \exp(\xi_{12} t) \exp(\beta_{12.2}\ b.age + \beta_{12.3}\ d.age),$$

it becomes apparent that a Gompertz baseline hazard is used for time t with parameters $\lambda_{12} > 0$ and ξ_{12}. An alternative would be to use a Weibull baseline hazard specified by

$$q_{12}(t) = \lambda_{12} \tau_{12} t^{\tau_{12}-1} \exp(\beta_{12.2}\ b.age + \beta_{12.3}\ d.age),$$

for $\lambda_{12}, \tau_{12} > 0$. These specifications and others will be discussed in Chapters 2, 3, and 4.

1.4 Overview of methods and literature

The multi-state survival models in this book are based upon the theory on *stochastic processes*. In probability theory, a stochastic process is a set of random variables representing the evolution of a process over time. More formally, the process is a collection of random variables $\{Y_t \mid t \in \mathcal{U}\}$, where $\mathcal{U}$ is some index set. If $\mathcal{U} = \{0, 1, \ldots\}$, then the process is in discrete time. If $\mathcal{U} = (0, \infty)$, then the process is in continuous time. Variable Y_t is the *state* of the process at time t. The theory on basic stochastic processes is well established and can be found in many textbooks; see, for example, Cox and Miller (1965), Norris (1997), Ross (2010), and Kulkarni (2011).

In probability theory, probabilistic properties are defined and are used to derive mathematical insight in the behaviour of a process over time. Statistics is primarily an applied branch of mathematics that uses definitions and results from probability theory to describe, understand, and predict processes in the material world. To be able to do this, statistics needs data and methods for distributional inference.

This book is concerned with the statistical modelling and estimation of multi-state processes. The most basic multi-state survival process is the standard survival model which consists of two states, alive and dead, and where there is only one event of interest, namely, death. Textbooks on this topic are, for example, Cox and Oakes (1984), Kalbfleisch and Prentice (2002), Collett (2003), and Aalen et al. (2008). The survival model will serve in Chapter 2 as a starting point for the statistical methods for more complex multi-state processes in Chapters 3 and 4. In multi-state models other than the standard survival model there is more than one event of interest. In a two-state model, for example, it is possible that not only the event of moving from state 1 to state 2 is of interest, but also the event of a transition back to state 1. Many processes in the material world can be described by multi-state models. The focus of this book is on biostatistics and on interval-censored data for continuous-time processes which include one absorbing state, typically the dead state.

Important methodological work on continuous-time multi-state models in the presence of interval censoring is presented in Kalbfleisch and Lawless (1985), Kay (1986), and Satten and Longini (1996). Jackson (2011) presents the freely available R package msm that provides a flexible framework for fitting a wide range of continuous-time multi-state models. Many textbooks on survival analysis contain extended material on multi-state models; see, for example, Hougaard (2000), Aalen et al. (2008), and Crowder (2012). The journal *Statistical Methods in Medical Research* devoted a whole issue to multi-state models; see Andersen (2002).

This book provides an overview of methods for multi-state survival models for interval-censored data, and extends the current methods by exploring and estimating time-dependent hazard models for transition intensities. A very general scoring algorithm for maximum likelihood estimation is presented in Section 4.6. Some of the material in this book is based on published work; see, for example, Van den Hout and Matthews (2008), Van den Hout and Matthews (2009), Van den Hout and Matthews (2010), Van den Hout and Tom (2013), and Van den Hout et al. (2014).

There are also topics related to multi-state processes which will not be discussed in depth in this book. One of these topics is the case where there is no interval censoring and exact times are available for all transitions in a continuous-time multi-state process. For modelling this kind of data, methods based upon counting processes can be used; see, for example, Aalen et al. (2008) and Putter et al. (2007) for a review of these methods and further references.

Attention to discrete-time multi-state processes is also limited. Section 4.1 defines these processes formally, but mainly as a contrast to continuous-time processes. In Section 8.1, the discrete-time process is used to approximate a continuous-time process.

For the longitudinal data, we assume that the sampling is independent of the history of the events of interest. Typically, this implies a prospective study where a random sample is taken from the population and observed subsequently. This independence assumption does not hold for retrospective studies. For the latter, see Kalbfleisch and Lawless (1988) for details and methods.

Two recent books on multi-state models are Beyersmann et al. (2012) and Willekens (2014). The former is on methods for longitudinal data with known transition times—the analysis of interval-censored data is not covered explicitly. The latter discusses interval-censored data to some extent using the term *status data*, and has a strong focus on how to use existing software for multi-state models. Unavoidably, there is some overlap between these two books and the current book. Specific for the current book, however, is the focus on interval-censored data, and the aim for a wider scope of the statistical modelling. This aim is realised by using a very general definition of the multi-state model which is unrestricted with respect to the number of states *and* with respect to possible (forward and backward) transitions. In addition, the current book discusses methods for Bayesian inference and frailty models, and includes a number of further topics of interest such as misclassification of state and missing data.

1.5 Data used in this book

The following is an overview of the data used in this book. Data are used for illustration purposes and the statistical analyses in this book should not be used in isolation to inform clinical practice.

- The cardiac allograft vasculopathy (CAV) data were collected at Papworth Hospital in the United Kingdom and are included in the R package msm (Jackson, 2011). Permission to use the data was given by Steven Tsui at Papworth Hospital. The data are introduced in Section 1.3.1. For up to 18 years, annual examinations for CAV took place after heart transplant. There is variation in time between examinations: it was not always exactly 1 year and some patients skipped one or more scheduled examinations. States are defined according to CAV status and death, and transitions between the living states are interval-censored. The number of individuals in the sample is 622.

- The Parkinson's disease data are provided by Dag Aarsland at the Norwegian Centre for Movement Disorders, Stavanger University Hospital, Norway. The data are introduced in Section 3.7.1. Individuals with Parkinson's disease were followed up between 1993 and 2005. Of interest are survival and the onset of dementia, which is interval-censored or right-censored. The number of individuals in the sample is 233.
- Longitudinal data from the English Longitudinal Study of Ageing (ELSA) are introduced in Section 4.8.1. The ELSA baseline is a representative sample of the English population aged 50 and older. ELSA contains information on health, economic position, and quality of life. Data from ELSA can be obtained via the UK Economic and Social Data Service. Follow-up data are used from 1998 up to 2009. The total number of individuals at baseline is 19,834. This book uses a random subsample of 1,000 individuals to illustrate the methods.
- The Medical Research Council Cognitive Function and Ageing Study (MRC CFAS) is a population-based longitudinal study of cognition and health conducted between 1991 and 2004 in the older population of England and Wales. The data are introduced in Section 4.10.1. Data were collected in two rural areas and three cities (Cambridgeshire, Nottingham, Gwynedd, Newcastle, and Oxford, respectively), with a sample size of at least 2,500 individuals at each site. In this book, the data from Newcastle are used with 2,512 individuals at baseline.
- Dutch data on body mass index (BMI) are used in Section 8.2.1. The data are cross-sectional, and men in the age range 15 up to 40 years old are classified in three states according to their BMI. Permission to use the data for this book was given by Jan van de Kassteele at the National Institute for Public Health and the Environment, RIVM, in Bilthoven, the Netherlands. The data used in this book are a subset of the data published in Van de Kassteele et al. (2012).

Chapter 2

Modelling Survival Data

2.1 Features of survival data and basic terminology

This chapter discusses statistical methods for parametric univariate survival models. The presentation is limited to methods that are relevant for the subsequent chapters on multi-state survival models. The reader who wants to know more about the univariate survival model is referred to textbooks on this topic such as Cox and Oakes (1984), Kalbfleisch and Prentice (2002), Collett (2003), and Aalen et al. (2008).

In the univariate survival model there is only one event of interest. In biostatistics, the event will often be death, or the onset of a condition or a disease. This is different from a multi-state survival model, which can be seen as a multivariate survival model with more than one event of interest. Univariate survival models can be found in many areas in statistics and—as a result—various names are used for the observed times to the event: *duration data*, *time-to-event data*, *failure-time data*, and *survival data*.

Censoring is the distinguishing feature of survival data. When an event is right-censored it did not occur during the study time. When the event is interval-censored it occurred in a time interval with known lower and upper bounds but at an unknown time within the interval. Left censoring is a specific form of interval censoring, namely, the case where the lower bound is zero. The models in this chapter assume that the censoring is non-informative; that is, the censoring mechanism is assumed to be independent of the stochastic time-to-event process. An example of informative censoring is an event time which is censored because the treatment associated with the event stopped due to a deterioration of the physical condition of the patient (Collett, 2003).

When an event is left-censored, it is at least known that the event took place. Left truncation (or delayed entry) refers to the situation where a process is unknown because the event time is less than some threshold.

The following example is from Kalbfleisch and Prentice (2002). Assume that the time origin is time of birth and that it is of interest whether people were exposed to some environmental risk. If individuals are selected for the study after responding to an advertisement, then individuals who died prior to the advertisement are not observed and data are left-truncated.

The time scale in a survival model is important. The time origin specifies from which moment survival is measured. Typical choices of the time scale are time from randomisation in a clinical trial, time from exposure in a study of infectious disease, or age in an observational study; see, for example, Bull and Spiegelhalter (1997) and Korn et al. (1997).

Model formulation in this chapter is with respect to the hazard, and inference is within the framework of Cox and Oakes (1984) and Kalbfleisch and Prentice (2002). A more rigorous mathematical approach for inference is based on counting processes and martingale theory (Andersen et al., 1993; Aalen et al., 2008).

An alternative to modelling the hazard is direct modelling of the survival time. When the logarithm of the survival time is used, this is called an accelerated failure time model; see, for example, Wei (1992) or Collett (2003).

The notation in this chapter is conventional. Random variable T denotes time to the event. For a value of T denoted t, the chapter starts with specifications of the density function $f(t)$, the hazard function $h(t)$, and the survivor function $S(t)$. Next maximum likelihood estimation is explained with special attention to piecewise-constant hazard models. Finally, an illustrative data analysis is presented.

2.2 Hazard, density, and survivor function

Distributions for time to event can be characterised by either the hazard function, the distribution function, the density function, or the survivor function. The relation between these four functions is as follows. For a continuous random variable T with sample space $[0,\infty)$, the distribution function is given by

$$F(t) = P(T \leq t) = \int_0^t f(u)du,$$

where f is the density function. The survivor function is

$$S(t) = 1 - F(t) = P(T > t).$$

The hazard is defined as

$$h(t) = \lim_{\Delta \downarrow 0} \frac{P(t \leq T \leq t + \Delta | T > t)}{\Delta}.$$

It follows that

$$h(t) = \frac{f(t)}{S(t)} = -\frac{\partial}{\partial t} \log\left(S(t)\right)$$

and

$$S(t) = \exp\left(-\int_0^t h(u)du\right).$$

Given the definition of the cumulative hazard function as $H(t) = \int_0^t h(u)du$, we can write $S(t) = \exp\left(-H(t)\right)$. In the context of interval censoring, note that

$$P(t_1 \leq T \leq t_2) = F(t_2) - F(t_1) = \int_{t_1}^{t_2} f(u)du = \int_{t_1}^{t_2} h(u)S(u)du.$$

Most of the multi-state survival analyses in this book are based upon parametric models for the hazard. Model validation, however, is partly based upon using non-parametric Kaplan–Meier estimates of survivor functions. To introduce the Kaplan–Meier estimate, first note that in the absence of censoring, the empirical survivor function is

$$\widehat{S}(t) = \frac{\text{Number of events} \leq t}{\text{Number of sample units in the data}}.$$

The following summarises the discussion of the Kaplan–Meier estimate in Collett (2003, Chapter 2). Say there are N sample units with observation times $t_1, ..., t_N$ which are either event times or censored event times. Say there are $r \leq N$ event times ordered as $t_{(1)}, ..., t_{(r)}$. Let n_j denote the number of sample units with an observation time at or after $t_{(j)}$, and let d_j be the units with an event time at $t_{(j)}$. Since there are n_j units without the event just before $t_{(j)}$ and d_j events at $t_{(j)}$, the probability of no event during $(t_{(j)} - \Delta,\ t_{(j)})$, for small $\Delta > 0$, is estimated by $1 - d_j/n_j = (n_j - d_j)/n_j$.

Since there are no events during $(t_{(j)},\ t_{(j+1)} - \Delta)$, the probability of no event during $(t_{(j)}, t_{(j+1)})$ is estimated by $(n_j - d_j)/n_j$ by taking the limit $\Delta \to 0$. Assuming independence of the events, the Kaplan–Meier estimate is given by

$$\widehat{S}(t) = \prod_{j=1}^{k} \frac{n_j - d_j}{n_j},$$

for $t_{(k)} \leq t < t_{(k+1)}$, $k = 1, 2, ..r$, with $\widehat{S}(t) = 1$ for $t < t_{(1)}$ and $t_{(k+1)} = \infty$. In absence of censoring, $\widehat{S}(t)$ reduces to the empirical survivor function.

The variance of $\widehat{S}(t)$ can be estimated by Greenwood's formula:

$$\widehat{Var}(\widehat{S}(t)) = \left(\widehat{S}(t)\right)^2 \sum_{j=1}^{k} \frac{d_j}{n_j(n_j - d_j)},$$

which is based upon a Taylor series approximation to $Var(\widehat{S}(t))$; see Collett (2003) for details.

Similar expressions are also available for the estimate of the cumulative hazard function $H(t)$; see, for example, Venables and Ripley (2002, Chapter 13). There are several options for constructing confidence intervals for $\widehat{S}(t)$, some of which use the estimated variance of $\widehat{H}(t)$. When using a software package to estimate confidence intervals for the survivor function, it is worthwhile to check which of the options is used.

2.3 Parametric distributions for time to event

2.3.1 Exponential distribution

A continuous random variable T has an exponential distribution with parameter $\lambda > 0$ if the probability density function is given by

$$f(t) = \begin{cases} \lambda \exp(-\lambda t) & t \geq 0 \\ 0 & t < 0. \end{cases}$$

This is denoted $T \sim Exp(\lambda)$. It follows that T has distribution function $F(t) = P(T \leq t) = 1 - \exp(-\lambda t)$ for $t \geq 0$. The survivor function is therefore given by $S(t) = P(T > t) = \exp(-\lambda t)$ for $t \geq 0$. The hazard function of T is

$$h(t) = \lambda \quad \text{for} \quad t \geq 0,$$

which follows directly from $h(t) = -\partial/\partial t \log\left(S(t)\right)$.

Mean survival is given by $\mathbb{E}(T) = 1/\lambda$, which illustrates that smaller values of $\lambda > 0$ correspond to better survival. The variance is $\text{Var}(T) = 1/\lambda^2$. The exponential distribution is memoryless in the sense that

$$P(T > u + t | T > u) = P(T > t),$$

for all $u, t > 0$. In terms of survival this means that given survival up to time u, survival after time u is independent of u.

2.3.2 *Weibull distribution*

A continuous random variable T has a Weibull distribution with scale parameter $\lambda > 0$ and shape parameter $\tau > 0$ if the probability density function is given by

$$f(t) = \begin{cases} \lambda \tau t^{\tau-1} \exp(-\lambda t^{\tau}) & t \geq 0 \\ 0 & t < 0. \end{cases}$$

This is denoted $T \sim Weibull(\lambda, \tau)$. It follows that T has distribution function $F(t) = P(T \leq t) = 1 - \exp(-\lambda t^{\tau})$ and survivor function $S(t) = P(T > t) = \exp(-\lambda t^{\tau})$, for $t \geq 0$. The hazard function is

$$h(t) = \lambda \tau t^{\tau-1} \quad \text{for} \quad t \geq 0.$$

The expected value is $\mathbb{E}[T] = \lambda^{-1/\tau}\Gamma(1+\tau^{-1})$ and the variance is

$$\text{Var}(T) = \lambda^{-2/\tau}\left(\Gamma(1+2/\tau) - \Gamma(1+1/\tau)^2\right),$$

where Γ is the gamma function; see, for example, Johnson et al. (1994, Chapter 21) and Hougaard (2000, Section 2.2).

Given survival up to time $u > 0$, the Weibull distribution is $S(t|u) = \exp\big(-\lambda(t^{\tau} - u^{\tau})\big)$. This can be used to simulate conditional Weibull survival times by the inversion method. For survival conditional on survival up to time u, define

$$T_{|u} = \exp\left(\frac{1}{\tau}\log\left(-\frac{1}{\lambda}\log(X) + u^{\tau}\right)\right), \tag{2.1}$$

for $X \sim U(0,1)$. First simulate a value x from the uniform distribution $U(0,1)$, and then plug this value into (2.1) to obtain simulated conditional survival time $t_{|u}$.

Note that simulating a conditional event time can be done using the algorithm for simulating a conditional survival time. For an event time, we have to replace X in (2.1) by $1-X$. Since $X \sim U(0,1)$ implies $1-X \sim U(0,1)$, the same algorithm can be used. An implementation of the algorithm can easily be checked by setting $u = 0$ and comparing sample mean and sample variance to the theoretical mean and variance.

2.3.3 *Gompertz distribution*

A continuous random variable T has a Gompertz distribution with parameters $\lambda > 0$ and ξ if the probability density function is given by

$$f(t) = \begin{cases} \lambda \exp(\xi t) \exp\left(-\lambda \xi^{-1}(\exp(\xi t) - 1)\right) & t \geq 0 \\ 0 & t < 0. \end{cases}$$

This is denoted $T \sim Gompertz(\lambda, \xi)$. It follows that T has distribution function $F(t) = P(T \leq t) = 1 - \exp\big(-\lambda\xi^{-1}(\exp(\xi t) - 1)\big)$ and survivor function $S(t) = P(T > t) = \exp\big(-\lambda\xi^{-1}(\exp(\xi t) - 1)\big)$, for $t \geq 0$. The hazard function is

$$h(t) = \lambda \exp(\xi t) \quad \text{for} \quad t \geq 0,$$

and $H(t) = \lambda/\xi\big(\exp(\xi t) - 1\big)$.

There is no formal restriction $\xi > 0$. However, if $\xi < 0$, then for t very large, the survivor function goes to $\exp(\lambda\xi^{-1}) > 0$, which implies that the event does not occur for a proportion of the population.

There are no closed-form expressions for the mean and the variance of the Gompertz distribution; see Grag et al. (1970), Pollard and Valkovics (1992), and Lenart (2014) for details and approximations.

A truncated Gompertz distribution is again Gompertz. Considering survival conditional on $T > u$, it follows that

$$\begin{aligned} P(T > t | T > u) &= \frac{S(t)}{S(u)} = \exp\Big(-\frac{\lambda}{\xi}\big(\exp(\xi t) - \exp(\xi u)\big)\Big) && (2.2) \\ &= \exp\Big(-\frac{\lambda \exp(\xi u)}{\xi}\big(\exp(\xi(t-u)) - 1\big)\Big). && (2.3) \end{aligned}$$

Equation (2.3) shows that $T - u \sim Gompertz\big(\lambda \exp(\xi u), \xi\big)$. Equation (2.2) can be used to simulate conditional Gompertz survival times. Define

$$T_{|u} = \frac{\log\big(\lambda \exp(\xi u) - \xi \log(X)\big) - \log(\lambda)}{\xi}, \tag{2.4}$$

for $X \sim U(0,1)$. As in the Weibull case, plug in simulated value x to get simulated conditional survival time $t_{|u}$; see also Bender et al. (2005).

2.3.4 *Comparing exponential, Weibull and Gompertz*

When it comes to modelling survival data, the distributions for time to event are best compared with respect to the change of the hazard over time. Summarising, we have

$$\begin{aligned} \text{exponential: } h(t) &= \lambda \\ \text{Weibull: } h(t) &= \lambda\tau t^{\tau-1} \\ \text{Gompertz: } h(t) &= \lambda \exp(\xi t) \end{aligned} \tag{2.5}$$

for $t \geq 0$.

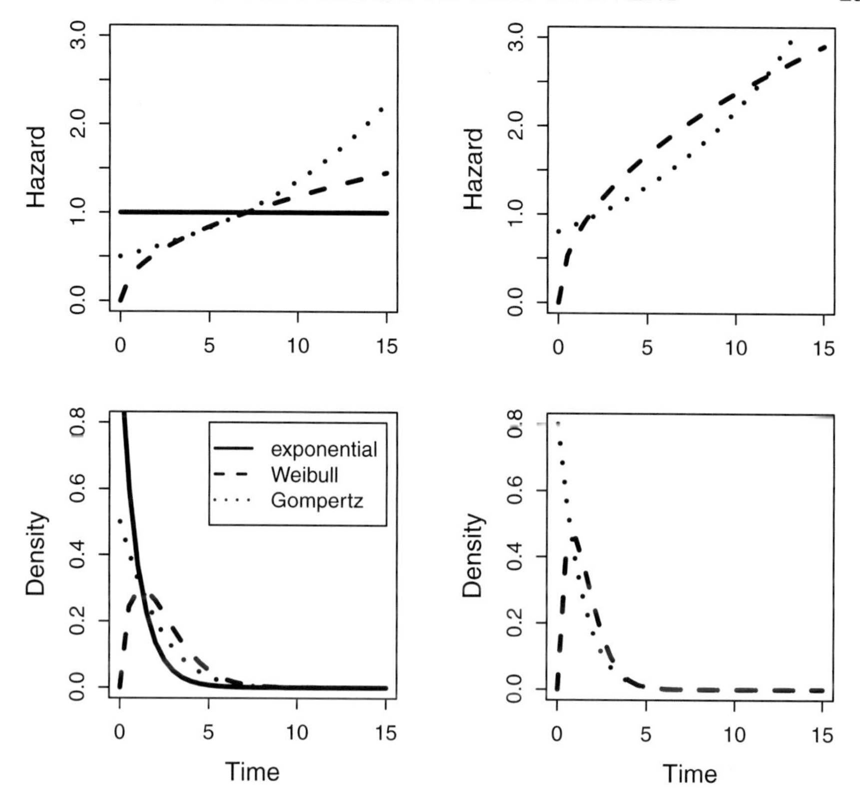

Figure 2.1 *Examples of hazards (top) and corresponding densities (bottom). At the left, Exp(1), Weibull(1/4, 3/2), and Gompertz(1/2, 1/10). At the right, Weibull(1/2, 3/2) and Gompertz(4/5, 1/10).*

The hazard is chosen for further statistical modelling because it does not change when conditioning on $T > u$. Consider the survivor function at time t conditional on survival up to u, where $t > u$. This conditional probability is given by

$$S(t|u) = P(T > t|T > u) = \frac{P(T > t, T > u)}{P(T > u)} = \frac{S(t)}{S(u)}.$$

Likewise, it holds that the conditional density is given by $f(t|u) = f(t)\,(S(u))^{-1}$. So both the survivor function and the density change in the presence of left truncation. This is not the case with the hazard. The hazard at time t is independent from survival up to any time before t.

Figure 2.1 shows hazards and corresponding densities for the different distributions. In biostatistics, a constant hazard will often be too strong an assumption in the context of disease or mortality. However, the exponential distribution is a good starting point as it is the building block for more complex distributions. For example, the Weibull distribution is an extension of the exponential since the latter can be obtained by setting $\tau = 1$. Likewise, if $\xi = 0$ in the Gompertz distribution, then the resulting distribution is exponential.

2.4 Regression models for the hazard

Often there will be information available in addition to the event time or the censoring time. If this information is available as covariate values, then we can extend the statistical modelling by defining a regression model for the hazard. The general form for time-independent covariates is given by

$$h(t|\mathbf{x}) = h_0(t)\exp\left(\boldsymbol{\beta}^\top \mathbf{x}\right), \tag{2.6}$$

where $\boldsymbol{\beta} = (\beta_1, ..., \beta_p)^\top$ is a parameter vector, $\mathbf{x} = (x_1, ..., x_p)^\top$ is the covariate vector without an intercept, and $h_0(t)$ is the function which describes how the hazard depends on time $t \geq 0$. The function $h_0(t)$, which is called the *baseline hazard*, can be specified by the parametric distributions given in (2.5).

2.5 Piecewise-constant hazard

Hazard models can be specified where the hazard is constant within specified time intervals but is allowed to change between the intervals. Let time be subdivided into equidistant intervals by defining $u_1 = 0$, and $u_j = u_{j-1} + h$, for $j = 2, ..., J$, where $h > 0$ and J are specified depending on the application. We discuss two ways to define piecewise-constant hazards.

Firstly, hazard parameters can be specified separately for each time interval. Assume, for example, that $J = 3$ and that for each interval the hazard is specified by a separate parameter, then

$$h(t) = \begin{cases} \lambda_1 & t \in (u_1, u_2] \\ \lambda_2 & t \in (u_2, u_3]. \end{cases}$$

More in general for any J, define $h(t) = \sum_{j=1}^{J-1} \lambda_j \mathbb{1}(u_j < t \leq u_{j+1})$, where $\mathbb{1}(A)$ is equal to 1 if A is true and 0 otherwise. This will be called a *piecewise-constant hazard model*. To be able to estimate the $J-1$ parameters, data should be available for all intervals $(u_1, u_2], ..., (u_{J-1}, u_J]$.

A second way to define piecewise-constant hazards is to keep the hazards constant within specified time intervals and to allowed them to change between the intervals according to a parametric shape. This will be called a *piecewise-constant approximation*. The advantage is that there is no need to specify interval-specific hazard parameters: only the parameters for the parametric shape have to be defined. When there are many time intervals, this approach induces a parsimonious piecewise-constant approximation to a parametric model. As an example, say we use the parametric shape of the Gompertz distribution. The model is then defined by $h(t) = \sum_{j=1}^{J-1} h_j(t) \mathbb{1}(u_j < t \leq u_{j+1})$, where $h_j(t) = \lambda \exp(\xi u_j)$. For this model, only the parameters λ and ξ have to be estimated.

A piecewise-constant hazard model can be used when the hazard cannot be described well by a parametric shape. By introducing a series of independent λ-parameters an irregular-shaped hazard can be approximated piecewise-constantly. The disadvantages are that determining the optimum number of λ-parameters is not easy and that prediction beyond the follow-up time (typically u_J) is problematic.

Given a parametric baseline hazard, a piecewise-constant approximation is only of interest in the presence of time-dependent covariates. Assume that a parametric baseline hazard $h_0(t)$ is specified and that $x(t)$ is a covariate with stochastic change over time. An example in biostatistics would be blood pressure in the context of death as the event of interest. If the model is $h(t) = h\big(t|x(t)\big) = h_0(t)\exp\big(\beta x(t)\big)$ and $h_0(t)$ and β are given, then the value of $h\big(t|x(t)\big)$ is only known for those times t at which $x(t)$ is known. In the presence of intermediate observation times before death or right censoring, the hazard model can be estimated by using the observation times as a time grid and by approximating the parametric shape for the hazard piecewise-constantly.

Using piecewise-constant values for a time-dependent covariate is of course an approximation to a change of values which will—in most applications—take place in continuous time.

2.6 Maximum likelihood estimation

If all event times are observed, the likelihood is the joint density of the event times. In the presence of right censoring, the likelihood is a product of densities and survivor functions.

For individual i, $i = 1, \ldots, N$, two times are defined: $t_{b.i}$ is the time of left truncation, and t_i is either the time to the event or the time of right censoring. Define $\delta_i = 1$ if the event time is observed, $\delta_i = 0$ if the event time is right-censored. If all model parameters are collected in the vector $\boldsymbol{\theta}$, then the likelihood is

$$L(\boldsymbol{\theta}) = L(\boldsymbol{\theta}|t_i, t_{b.i}, \mathbf{x}_i, i = 1, ..., N) = \prod_{i=1}^{N} S(t_i|t_{b.i}, \mathbf{x}_i)^{1-\delta_i} f(t_i|t_{b.i}, \mathbf{x}_i)^{\delta_i}$$
$$= \prod_{i=1}^{N} S(t_i|t_{b.i}, \mathbf{x}_i) h(t_i|\mathbf{x}_i)^{\delta_i}, \quad (2.7)$$

where the second equation follows from $f(t|u) = h(t)S(t|u)$ for $t \geq u \geq 0$; see Section 2.2. For the exponential hazard model without covariates and without left truncation, the maximum likelihood estimate is available in closed form; that is, $\widehat{\lambda} = r\left(\sum_{i=1}^{N} t_i\right)^{-1}$, where r is the total number of events. For the Weibull and the Gompertz hazard model, numerical maximisation is needed. More information on maximum likelihood estimation for the exponential and the Weibull model can be found in Collett (2003, Section 4.3.1). For the Gompertz model, see Grag et al. (1970).

Interval censoring can easily be taken into account. If for an individual i, the event time is known to lie in the interval $(t_{i1}, t_{i2}]$, then the likelihood contribution for that individual is $S(t_{i1}|t_{b.i}, \mathbf{x}_i) - S(t_{i2}|t_{b.i}, \mathbf{x}_i)$; see, for example, Sun (2006, Section 2.3). Right censoring can thus be seen as a special case of interval censoring, namely, the case where $t_{i2} = \infty$.

For a piecewise-constant hazard model, survival time is assessed for each interval in the time grid. Say the grid is given by $u_1 = 0, u_2, ..., u_J$, where u_J is equal to the largest observation time. For individual i, the relevant part of the grid is $u_1, u_2, ..., u_{J_i}$ when $t_i \in (u_{J_i-1}, u_{J_i}]$. The likelihood in absence of left truncation is given by

$$L(\boldsymbol{\theta}) = \prod_{i=1}^{N} \prod_{j=1}^{J_i} S_j(t_{ij}, \mathbf{x}_i) \left(h(u_j|\mathbf{x}_i)\right)^{\delta_{ij}}, \quad (2.8)$$

where $t_{ij} = \min(t_i, u_{j+1}) - \min(t_i, u_j)$, $\delta_{ij} = 1$ when death occurs in interval $(u_j, u_{j+1}]$ and zero otherwise, and S_j is the survivor function for the time spent in $(u_j, u_{j+1}]$; see Hougaard (2000, Section 2.2.4).

For a piecewise-constant approximation using the grid $u_1 = 0, u_2, ..., u_J$, the likelihood is given by (2.8) but the hazard function $h(u_j)$ underlying a likelihood contribution for interval $(u_j, u_{j+1}]$ is evaluated according to the imposed parametric shape; see Section 2.5.

2.7 Example: survival in the CAV study

This section illustrates the hazard models in this chapter. Special attention will be given to the piecewise-constant approximation of a parametric hazard model because understanding this approach is of importance for the multi-state models in the subsequent chapters.

The longitudinal study with angiographic examinations after heart transplant (Sharples et al., 2003) is introduced in Section 1.3, where a four-state survival model is defined for history of cardiac allograft vasculopathy (CAV); see Figure 1.4. The current section will investigate univariate survival models where the event of interest is death. Dichotomised CAV state (no CAV versus CAV) will be included in the survival model as a time-dependent covariate.

Using a parametric two-state survival model with CAV included as a time-dependent covariate allows investigating the effect of CAV on survival, but the model cannot be used for prediction without additional modelling of the stochastic process which underlies the onset of CAV. This is a disadvantage of the standard survival model. In contrast, Section 1.3 illustrates the advantage when using a parametric multi-state model for the CAV data: it allows predicting both the CAV process and survival.

For the two-state survival model, two time scales will be considered, namely, *years after heart transplant* and *age*. The former is the most relevant for the process of interest; the latter is used to illustrate left truncation and for purpose of comparison.

The number of individuals in the longitudinal study is $N = 622$ and, as can be deduced from the information given in Table 1.1, the number of individuals who died during the follow-up is 251. In the survival models with years after transplant as time scale, we use three covariates, namely, gender (*sex* with 0/1 for men/women), the age of the donor (*d.age*), and age at transplant (*b.age*). Hence the hazard regression model is given by

$$h(t|\mathbf{x}) = h_0(t)\exp\left(\beta_1\, sex + \beta_2\, d.age + \beta_3\, b.age\right), \tag{2.9}$$

where $h_0(t)$ is the baseline hazard. Investigated are the exponential baseline $h_0(t) = \exp(\beta_0)$, the Weibull baseline $h_0(t) = \exp(\beta_0)\tau t^{\tau-1}$, and the Gompertz baseline $h_0(t) = \exp(\beta_0 + \xi t)$. Results are presented in Table 2.1, where the Gompertz model is denoted by Gompertz I. According to Akaike's information criterion (AIC) introduced by Akaike (1974), the Gompertz model with the lowest AIC is the best of the three parametric choices; see Section 4.7 for the definition of AIC.

The estimated parameters of the Gompertz model show that the hazard of death increases with years after the transplant ($\widehat{\xi} > 0$), which is according to expectation. Furthermore, the transplant for women is associated with a higher risk compared to men ($\widehat{\beta_1} > 0$), and both increased age at baseline and increased age of the donor are positively associated with the hazard of death ($\widehat{\beta_2}, \widehat{\beta_3} > 0$).

Table 2.1 *Parameter estimates (and standard errors) for the models for survival in the CAV study, and Akaike's information criterion (AIC).*

	Exponential	Weibull	Gompertz I	Gompertz II
Time scale defined by years since transplant				
β_0	−3.552 (0.317)	−4.019 (0.368)	−5.319 (0.386)	−4.687 (0.377)
β_1	0.080 (0.203)	0.131 (0.204)	0.343 (0.205)	0.370 (0.202)
β_2	0.008 (0.006)	0.009 (0.006)	0.013 (0.006)	0.008 (0.006)
β_3	0.014 (0.006)	0.016 (0.006)	0.028 (0.007)	0.024 (0.007)
β_4				0.891 (0.146)
τ		1.159 (0.063)		
ξ			0.177 (0.016)	0.100 (0.023)
AIC	1846	1841	1742	
Time scale defined by age				
γ_0	−2.996 (0.178)	−9.946 (1.412)	−5.030 (0.404)	
γ_1	0.016 (0.201)	0.198 (0.203)	0.211 (0.204)	
γ_2	0.011 (0.005)	0.005 (0.006)	0.006 (0.006)	
τ		2.573 (0.328)		
ξ			0.042 (0.007)	
AIC	1849	1823	1819	

The estimation of the effects of the risk factors can be compared with the estimation of the semi-parametric Cox regression model (Cox, 1972). This will provide insight into the sensitivity of this estimation to the induced parametric shapes. The Cox model specification in the current context is

$$h^*(t|\mathbf{x}) \quad = \quad h_0^*(t)\exp\left(\beta_1^*\ sex + \beta_2^*\ d.age + \beta_3^*\ b.age\right)$$

This model is semi-parametric because the baseline hazard is non-parametric. For the details of this model and the estimation see Terneau and Grambsch (2000) and the R package survival. The effects of the risk factors are estimated at $\widehat{\beta}_1^* = 0.330$ (0.205), $\widehat{\beta}_2^* = 0.012$ (0.006), and $\widehat{\beta}_1^* = 0.029$ (0.007), with estimated standard errors in parentheses.

The estimated effects of the risk factors for the Cox model are similar to the estimated effects for the Gompertz I model in Table 2.1. A further comparison between the two models is shown in Figure 2.2 which depicts cumulative

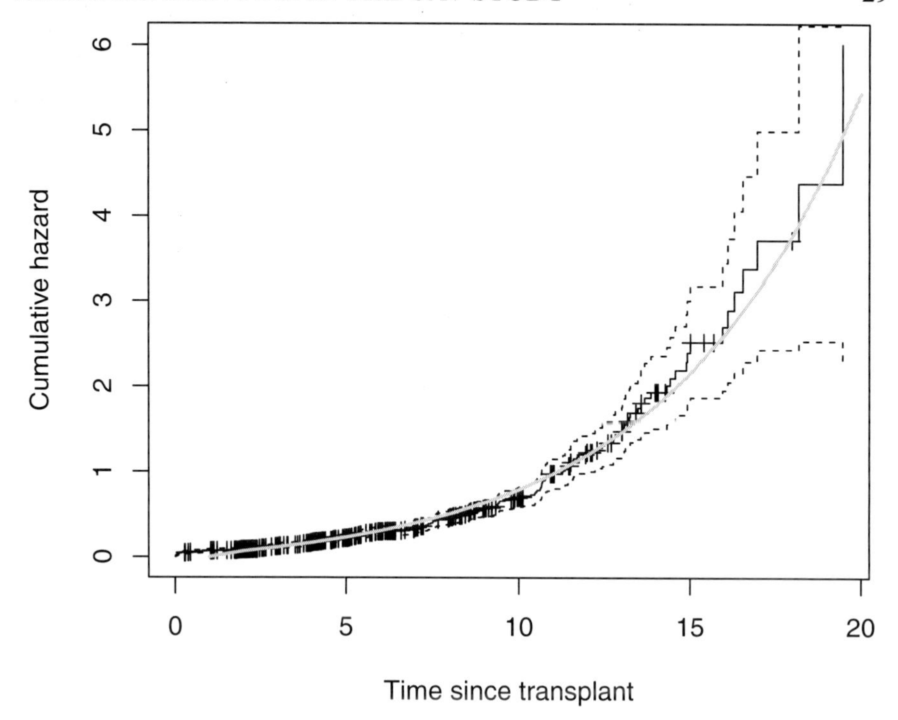

Figure 2.2 *Estimated cumulative hazard for survival in the CAV study. Black lines for Cox model (with 95% confidence intervals, and with plus signs for event times), and grey line for the Gompertz model. Both cumulative hazards are conditional on the means of the covariate values in the data.*

hazards. For the Cox model, Figure 2.2 shows the cumulative hazard conditional on the means of the covariate values in the data, and the corresponding 95% confidence interval. Also shown is the cumulative hazard derived from the Gompertz model (again for the mean of the covariate values). The two cumulative hazards are quite similar.

The comparison with the Cox model shows that for the years during the follow-up in the CAV study, the parametric Gompertz shape for the hazard seems to capture the time dependency quite well. If we are willing to assume that the time dependency beyond the follow-up of the study follows a parametric shape, then the comparison with the Cox model implies that Gompertz model is a good candidate for long-term prediction.

For the survival models with age as the time scale t, the hazard regression model is given by

$$h(t|\mathbf{x}) = h_0(t)\exp(\gamma_1\, sex + \gamma_2\, d.age)\,.$$

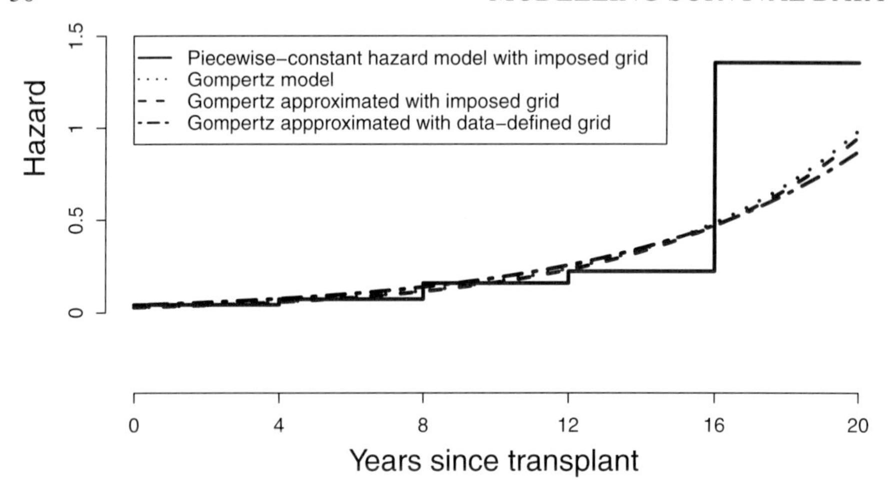

Figure 2.3 *Estimated change of hazard over time for CAV data. The piecewise-constant hazard model and the Gompertz model. For the latter, the parametric version and two piecewise-constant approximations of the parametric shape. Hazard depicted for a man 50 years of age with a transplant from a donor aged 28.*

Estimation is by maximising the likelihood function (2.7) where left truncation is taken into account because not all individuals have the transplant at the same age. Also for this time scale, the Gompertz model performs best according to the AIC; see Table 2.1 where the γ_0s are the counterparts of the β_0s in the models for year since transplant.

Going back to the time scale years after transplant, the estimation of the Gompertz model will be adapted so that CAV state can be added as a time-dependent covariate. First the piecewise-constant hazard model is investigated for the model without CAV state. For the grid $(u_1, ..., u_6) = (0, 4, 8, 12, 16, 20)$, Figure 2.3 shows the estimated piecewise-constant hazard model for a man fifty years of age with a transplant from a donor aged 28.

This can be compared to the piecewise-constant approximation of the Gompertz model. The likelihood for this approximation is given by (2.8), where the hazard function $h(u_j)$ underlying a likelihood contribution for interval $(u_j, u_{j+1}]$ is evaluated according to the Gompertz hazard halfway through the interval; that is, at time $t = (u_j + u_{j+1})/2$. Even though the grid $(u_1, ..., u_6)$ is rather coarse, Figure 2.3 shows that the estimated hazard is very similar to the Gompertz shape as estimated by using likelihood (2.7) for the Gompertz model.

When there are repeated observations for each individual, an alternative to using an imposed grid is to use individual-specific grids. For example, if individual i is observed at times t_{i1}, t_{i2}, t_{i3}, and t_{i4}, then define the grid for the likelihood contribution for i by $(t_{i1}, t_{i2}]$, $(t_{i2}, t_{i3}]$, and $(t_{i3}, t_{i4}]$. In this way, the grid is allowed to vary across individuals with various follow-up times. As shown in Figure 2.3, when this data-defined grid approach is used, the hazard is very similar to the estimation resulting without the piecewise-constant approximation.

Next, CAV state as binary covariate is added as time-dependent covariate to the Gompertz model. Covariate $cav(t)$ is defined by value 0 for no CAV at time t, and 1 otherwise. The model is now

$$h(t|\mathbf{x}) = h_0(t) \exp\left(\beta_1 \, sex + \beta_2 \, d.age + \beta_3 \, b.age + \beta_4 \, cav(t)\right). \tag{2.10}$$

CAV state is only observed at the follow-up times—stochastic behaviour between observation times is interval-censored. In this case, using individual-specific grids as defined by individual follow-up times is the obvious choice. Results are presented in Table 2.1 under Gompertz II. As expected, presence of CAV is associated with an increased hazard of death ($\widehat{\beta}_4 > 0$).

The estimation of the effects of the risk factors can again be compared with the estimation of the semi-parametric Cox regression model

$$h^*(t|\mathbf{x}) = h_0^*(t) \exp\left(\beta_1^* \, sex + \beta_2^* \, d.age + \beta_3^* \, b.age + \beta_4^* \, cav(t)\right),$$

with estimates $\widehat{\beta}_1^* = 0.406$ (0.206), $\widehat{\beta}_2^* = 0.007$ (0.006), $\widehat{\beta}_1^* = 0.030$ (0.007), and $\widehat{\beta}_4^* = 0.796$ (0.152). Also after adding the time-dependent covariate, differences are small between estimated effects of risk factors in the Gompertz model and the Cox model.

Although it is relatively easy to estimate a survival model which includes a stochastic time-dependent covariate, the use of the estimated model is limited. First off—as mentioned at the start of Chapter 1—the model cannot be used for prediction without additional modelling of the stochastic process which underlies the change of covariate values. Secondly, the model is no longer a proportional hazard model. As an example, consider the interpretation of the β_1 parameter in (2.9) for the effect of gender on the hazard of death. The ratio $h(t|sex = 1)/h(t|sex = 0) = \exp(\beta_1)$ can be used to quantify this effect conditional on fixed covariate values for $d.age$ and $b.age$. But this interpretation is not possible for model (2.10) where the time-dependent covariate induces a time-dependent hazard ratio; see also, for example, Collett (2003, Chapter 8).

Chapter 3

Progressive Three-State Survival Model

3.1 Features of multi-state data and basic terminology

In a multi-state model the events of interest are the transitions between the states. A multi-state model where one of the states is the dead state (or more generally, an absorbing state) is a multi-state survival model.

This chapter discusses the continuous-time progressive three-state survival model defined by two living states and a dead state. Chapter 4 will introduce the general multi-state model defined for any number of states. The current chapter is limited to the progressive three-state model. This model is one of the simplest extensions of the univariate survival model. In the absence of repeated transitions between two states, there is a relatively easy link between the transition intensities and the transition probabilities. In many applications the progressive three-state model is of primary interest, and readers who do not want to consider other models can choose to skip Chapter 4.

Features of multi-state data and basic terminology will be introduced here using an example of a progressive illness-death model, where state 1 is the healthy state, state 2 is the illness state, and state 3 is the dead state. Because the model is progressive, there are three transitions possible, namely, $1 \rightarrow 2$, $1 \rightarrow 3$, and $2 \rightarrow 3$. It is assumed that longitudinal data are available where individuals who are alive are classified in one of the two living states at prescheduled observation times, which are allowed to vary between individuals. Times of follow-up within individuals are not necessarily equidistant. It is also assumed that the time of death is always recorded if death occurs within the follow-up of the study.

With regard to statistical modelling, the primary quantities in the model are the transition-specific hazards. These hazards are also called *transition intensities* and each intensity represents at any given time the instantaneous risk of moving between two specified states. For the progressive three-state model there are three transition intensities corresponding to the three transitions. *Transition probabilities* represent the chance of a transition between two states within a specified time interval.

Censoring in a multi-state process is defined analogously to the censoring in the univariate survival process. An observed multi-state process is interval-censored when for a time interval $(L,R]$, a state is observed at L and at R, but state occupancies within $(L,R]$ are not observed. Right censoring is defined by $(L,\infty]$ and left censoring by $(0,R]$. Interval censoring for a multi-state process can have more severe consequences than for a one-event-only process. For the three-state progressive model, when state 1 is observed at L and state 2 at $R \neq L$, the exact time of the transition is interval-censored. However, it is still known that a transition took place. When state 1 is observed at L and state 3 at $R \neq L$, then the interval censoring is more severe: it is unknown whether the trajectory was $1 \rightarrow 2 \rightarrow 3$ or $1 \rightarrow 3$. When a model allows for backward transitions from state 2 to state 1, then even less is known as multiple transition times for states 1 and 2 could have been censored.

In biostatistics, transitions between living states in a multi-state model are often interval-censored since constant monitoring of the process is too expensive, or just not feasible. For example, if the illness in the progressive three-state model starts with an infection, then the onset of illness may be hard to pinpoint in time.

The choice of time scale is important and will vary according to the application. Specific to a multi-state process, more than one time scale can be defined. In the progressive three-state model, time can be defined by both time spent in state 1 and time spent in state 2. In addition, the time dependency can be modelled transition-specific. For example, time spent in state 1 may have a different effect on the risk of a transition to the dead state compared to the effect on the risk of a transition into the illness state.

Statistical analysis of multi-state survival processes can be found in many areas. Research in biostatistics that is related to work in this chapter is presented in Kalbfleisch and Lawless (1985), Kay (1986), and Jackson (2011). Theory for multi-state models is derived from theory on Markov processes; see, for example, Cox and Miller (1965) and Norris (1997).

The models in this chapter are parametric. An alternative is presented by Frydman and Szarek (2009), who present non-parametric estimation of a progressive three-state process in the presence of interval censoring. Yet another option is to define semi-parametric models using splines; this is discussed in a more general setting in Section 8.6.

3.2 Parametric models

Let t denote time in a progressive continuous-time three-state survival model. For the moment, we refrain from specifying a model for the hazards and use $q_{rs}(t)$ to denote the hazard for transition $r \to s$ at time t.

Given time interval $(t_1, t_2]$, the cumulative hazard functions for leaving state 1 and 2 are given by

$$H_1(t_1,t_2) = \int_{t_1}^{t_2} q_{12}(u) + q_{13}(u)du \qquad \text{and} \qquad H_2(t_1,t_2) = \int_{t_1}^{t_2} q_{23}(u)du,$$

respectively. Given that state 3 is an absorbing state and that the model is progressive, if follows that $q_{21}(t) = q_{31}(t) = q_{32}(t) = 0$. Let random variable $Y_t \in \{1,2,3\}$ denote the state at time $t \geq 0$. Transition probabilities $p_{rs}(t_1,t_2) = P(Y_{t_2} = s | Y_{t_1} = r)$ are given by

$$\begin{aligned}
p_{11}(t_1,t_2) &= \exp\left(-H_1(t_1,t_2)\right) \\
p_{12}(t_1,t_2) &= \int_{t_1}^{t_2} \exp\left(-H_1(t_1,u)\right) q_{12}(u) \exp\left(-H_2(u,t_2)\right) du \\
p_{13}(t_1,t_2) &= 1 - p_{11}(t_1,t_2) - p_{12}(t_1,t_2)
\end{aligned}$$

$$\begin{aligned}
p_{21}(t_1,t_2) &= 0 & p_{31}(t_1,t_2) &= 0 \\
p_{22}(t_1,t_2) &= \exp\left(-H_2(t_1,t_2)\right) & p_{32}(t_1,t_2) &= 0 \\
p_{23}(t_1,t_2) &= 1 - p_{22} & p_{33}(t_1,t_2) &= 1.
\end{aligned}$$

3.2.1 *Exponential model*

In the exponential model, transition-specific hazards are specified by constants. For the current three-state model, let $q_{rs}(t) = q_{rs}$ for all relevant states r and s at time t. It follows that

$$\begin{aligned}
p_{11}(t_1,t_2) &= \exp\left(-H_1(t_1,t_2)\right) = \exp\left(-(q_{12}+q_{13})(t_2-t_1)\right) \\
p_{22}(t_1,t_2) &= \exp\left(-H_2(t_1,t_2)\right) = \exp\left(-q_{23}(t_2-t_1)\right),
\end{aligned}$$

and

$$\begin{aligned}
p_{12}(t_1,t_2) &= \int_{t_1}^{t_2} \exp\left(-H_1(t_1,u)\right) q_{12} \exp\left(-H_2(u,t_2)\right) du \\
&= \exp\left((q_{12}+q_{13})t_1 - q_{23}t_2\right) \int_{t_1}^{t_2} q_{12} \exp\left(-(q_{12}+q_{13}-q_{23})u\right) du \\
&= \frac{q_{12}}{q_{12}+q_{13}-q_{23}} \Big(\exp\left(-q_{23}(t_2-t_1)\right) - \exp\left(-(q_{12}+q_{13})(t_2-t_1)\right) \Big),
\end{aligned}$$

for $q_{23} \neq q_{12} + q_{13}$. If $q_{23} = q_{12} + q_{13}$, then

$$p_{12}(t_1, t_2) = q_{12} \exp\big(-q_{23}(t_2 - t_1)\big)(t_2 - t_1).$$

The remaining transition probabilities can be derived directly. These formulas show that only elapsed time $t_2 - t_1$ is used to compute the transition probabilities. It is also possible to derive the above expressions for the transition probabilities using the theory on continuous-time processes; see Appendix A for details.

3.2.2 *Weibull model*

In the Weibull model, transition-specific hazards are time-dependent. For the current three-state model, let $q_{rs}(t) = \lambda_{rs}\tau_{rs}t^{\tau_{rs}-1}$, where $\lambda_{rs}, \tau_{rs} > 0$, for all relevant states r and s. We can derive

$$\begin{aligned} p_{11}(t_1, t_2) &= \exp\big(-H_1(t_1, t_2)\big) = \exp\Big(-\lambda_{12}\left(t_2^{\tau_{12}} - t_1^{\tau_{12}}\right) - \lambda_{13}\left(t_2^{\tau_{13}} - t_1^{\tau_{13}}\right)\Big) \\ p_{22}(t_1, t_2) &= \exp\big(-H_2(t_1, t_2)\big) = \exp\Big(-\lambda_{23}\left(t_2^{\tau_{23}} - t_1^{\tau_{23}}\right)\Big). \end{aligned}$$

Using the notation in Section 2.3.2, for $T \sim \mathit{Weibull}(\lambda_{23}, \tau_{23})$ it follows that $p_{22}(t_1, t_2) = P(T > t_2 | T > t_1) = S(t_2)/S(t_1)$.

For the transition from state 1 to state 2, it follows that

$$p_{12}(t_1, t_2) = \int_{t_1}^{t_2} \exp\big(-H_1(t_1, u)\big) q_{12}(u) \exp\big(-H_2(u, t_2)\big) du.$$

In general, the integral for $p_{12}(t_1, t_2)$ does not have a closed-form expression. However, the integrand can be written in closed form and one-dimensional numerical integration can be undertaken to approximate the integral for any t_1 and t_2.

For the restricted case where $\tau_{12} = \tau_{13} = \tau_{23} \stackrel{\mathrm{d}}{=} \tau$, the integral for $p_{12}(t_1, t_2)$ can be expressed in closed form:

$$p_{12}(t_1, t_2) = \\ \frac{\lambda_{12}}{\lambda_{12} + \lambda_{13} - \lambda_{23}} \Big(\exp\big(-\lambda_{23}(t_2^{\tau} - t_1^{\tau})\big) - \exp\big(-(\lambda_{12} + \lambda_{13})(t_2^{\tau} - t_1^{\tau})\big) \Big);$$

see Omar et al. (1995).

3.2.3 Gompertz model

In the Gompertz model, transition-specific hazards are time-dependent. For the current three-state model, let $q_{rs}(t) = \lambda_{rs}\exp(\xi_{rs}t)$, where $\lambda_{rs} > 0$, for all relevant states r and s.

$$
\begin{aligned}
p_{11}(t_1,t_2) &= \exp\left(-H_1(t_1,t_2)\right) \\
&= \exp\left(-\int_{t_1}^{t_2} \lambda_{12}\exp(\xi_{12}u) + \lambda_{13}\exp(\xi_{13}u)du\right) \\
&= \exp\left(-\left(\frac{\lambda_{12}}{\xi_{12}}(\exp(\xi_{12}t_2) - \exp(\xi_{12}t_1)) + \frac{\lambda_{13}}{\xi_{13}}(\exp(\xi_{13}t_2) - \exp(\xi_{13}t_1))\right)\right)
\end{aligned}
$$

$$
p_{22}(t_1,t_2) = \exp\left(-H_2(t_1,t_2)\right) = \exp\left(-\frac{\lambda_{23}}{\xi_{23}}(\exp(\xi_{23}t_2) - \exp(\xi_{23}t_1))\right)
$$

and

$$
p_{12}(t_1,t_2) = \int_{t_1}^{t_2} \exp\left(-H_1(t_1,u)\right)\lambda_{12}\exp(\xi_{12}u)\exp\left(-H_2(u,t_2)\right)du.
$$

Probability $p_{22}(t_1,t_2)$ is equal to $P(T > t_2 | T > t_1)$ for $T \sim Gompertz(\lambda_{23}, \xi_{23})$.

The above expressions only hold for $\xi_{rs} \neq 0$. If $\xi_{rs} = 0$ for given (r,s), then the hazard for the transition $r \to s$ is an exponential hazard and the expressions can easily be adapted.

The integral for $p_{12}(t_1,t_2)$ does not have a closed-form expression. However, the integrand can be written in closed form and one-dimensional numerical integration can be undertaken. Because of the exponential functions evaluated for time t, re-scaling t can help to prevent numerical overflow in the implementation of this model.

3.2.4 Hybrid models

It is possible to define hybrid models where parametric shapes are transition specific. An example is the three-state progressive model with Gompertz hazards for death and a Weibull hazard for the transition to state 2. Derivation of transition probabilities is similar to the above. The cumulative hazard function for leaving state 1 is

$$
\begin{aligned}
H_1(t_1,t_2) &= \int_{t_1}^{t_2} \lambda_{12}\tau_{12}u^{\tau_{12}-1} + \lambda_{13}\exp(\xi_{13}u)du \\
&= \lambda_{12}t_2^{\tau_{12}} + \frac{\lambda_{13}}{\xi_{13}}\exp(\xi_{13}t_2) - \lambda_{12}t_1^{\tau_{12}} - \frac{\lambda_{13}}{\xi_{13}}\exp(\xi_{13}t_1).
\end{aligned}
$$

The cumulative hazard function $H_2(t_1,t_2)$ for leaving state 2 is derived from the Gompertz hazard with λ_{23} and ξ_{23}. Transition probabilities for any time interval $(t_1,t_2]$ can be derived using these two cumulative hazard functions.

3.3 Regression models for the hazards

Analogously to the univariate survival model, statistical modelling can be extended by defining regression models for the transition-specific hazards. For the transition $r \to s$, the general form for time-independent covariates is given by

$$q_{rs}(t|\mathbf{x}) \quad = \quad q_{rs.0}(t)\exp\left(\boldsymbol{\beta}_{rs}^{\top}\mathbf{x}\right), \tag{3.1}$$

where $\boldsymbol{\beta}_{rs} = (\beta_{rs.1},...,\beta_{rs.p})^{\top}$ is a parameter vector, and $\mathbf{x} = (x_1,...,x_p)^{\top}$ is the covariate vector without an intercept. The function $q_{rs.0}(t)$ describes how the hazard depends on time $t > 0$ and will be called the baseline hazard. Examples of parametric choices for the baseline hazard are the Weibull and the Gompertz hazards as described above.

3.4 Piecewise-Constant hazards

Analogously to the univariate survival model, we distinguish *piecewise-constant hazard models* from *piecewise-constant approximations* of parametric shapes. In both approaches, the hazard is constant within specified time intervals but is allowed to change between the intervals. Let time be subdivided into equidistant intervals by defining $u_1 = 0$, and $u_j = u_{j-1} + h$, for $j = 2,...,J$, where $h > 0$ and J are specified depending on the application.

In a piecewise-constant hazard model a transition-specific hazard parameter is specified separately for each time interval. For example, if $q_{rs}(t) = \sum_{j=1}^{J-1} q_{rs.j}\mathbb{1}(u_j < t \leq u_{j+1})$ for transition $r \to s$, then there are $J-1$ parameters for this transition.

In a piecewise-constant approximation, the transition-specific hazards are constant within the time intervals and change between time intervals according to a parametric shape. When there are many time intervals, this will define a parsimonious piecewise-constant approximation to a parametric model. For the Gompertz model, for example, define $q_{rs}(t) = \sum_{j=1}^{J-1} q_{rs.j}(t)\mathbb{1}(u_j < t \leq u_{j+1})$, where $q_{rs.j}(t) = \lambda_{rs}\exp(\xi_{rs}u_j)$.

Piecewise-constant hazards lead to computational advantages because constant hazards induce closed-form expressions for the transition probabilities; see the equations for the exponential model. Using the closed forms will lead to faster estimation since there are no integrals to approximate.

The piecewise-constant hazard model can be used when a hazard cannot be described well by a parametric shape. A disadvantage is that the number of parameters increase quickly with the increase of number of transition-specific hazards in the multi-state model.

The piecewise-constant approximation of a parametric hazard allows the computational advantage and makes it possible to predict the process beyond the time of follow-up. An additional advantage of the piecewise-constant approximation is that time-dependent covariates can be taken into account relatively easily when the grid is defined by the data at hand. Assume that a parametric baseline hazard $q_{rs}(t)$ is specified and that $x(t)$ is a covariate with stochastic change over time. In the presence of interval censoring the values of $x(t)$ are only known at the observation times. The hazard model can be estimated by using individual-specific time grids for the likelihood contributions. This approximation was also used in the univariate model; see Sections 2.5 and 2.7.

In a piecewise-constant hazard model or in a piecewise-constant approximation with imposed grid, an observed time interval may consist of more than one subinterval with constant hazards. To illustrate this with an example, say the grid is $(u_1, u_2, u_3) = (0, 2, 4)$ and the time interval is $(t_1 = 1, t_2 = 3]$. The transition matrix for $(t_1, t_2]$, is given by

$$\mathbf{P}(t_1,t_2) = \mathbf{P}(t_1,t_2|\mathbf{x}) = \begin{pmatrix} p_{11}(t_1,t_2) & p_{12}(t_1,t_2) & p_{13}(t_1,t_2) \\ 0 & p_{22}(t_1,t_2) & p_{23}(t_1,t_2) \\ 0 & 0 & 1 \end{pmatrix},$$

where $p_{rs}(t_1,t_2) = P(Y(t_2) = s|Y(t_1) = r, \mathbf{x})$. It holds that $\mathbf{P}(t_1,t_2) = \mathbf{P}(t_1,u_2)\mathbf{P}(u_2,t_2)$, so $\mathbf{P}(t_1,t_2)$ can be approximated piecewise-constantly by computing both matrices at the right-hand side using constant hazards.

One of the first publications on using piecewise-constant intensities to deal with time dependency in continuous-time models is Faddy (1976). Additional discussions and models can be found in Kalbfleisch and Lawless (1985) and Kay (1986).

3.5 Maximum likelihood estimation

Estimation of model parameters is undertaken by maximising the log-likelihood function. First, we describe the likelihood contribution for a single individual i with observation times $t_1, ..., t_J$, where the state at t_J is allowed to be right-censored. In many applications integer J varies across individuals, but to simplify notation the index i is suppressed. Let $\mathbf{y}$ denote the observed trajectory defined by the series of states $y_1, ..., y_J$ corresponding to the times

$t_1, ..., t_J$. Let $\mathbf{x}$ denote the vector with the covariate values. Using the Markov assumption, the contribution of the individual to the likelihood is given by

$$\begin{aligned} L_i(\boldsymbol{\theta}|\mathbf{y},\mathbf{x}) &= P(Y_2 = y_2, ..., Y_J = y_J | Y_1 = y_1, \mathbf{x}) \\ &= \left(\prod_{j=2}^{J-1} P(Y_j = y_j | Y_{j-1} = y_{j-1}, \mathbf{x}) \right) C(y_J | y_{J-1}, \mathbf{x}), \quad (3.2) \end{aligned}$$

where $\boldsymbol{\theta}$ is the vector with the model parameters. The multiplicative term $C(y_J|y_{J-1},\mathbf{x})$ is defined as follows. If a living state at t_J is observed, then

$$C(y_J|y_{J-1},\mathbf{x}) = P(Y_J = y_J|Y_{J-1} = y_{J-1},\mathbf{x}).$$

If the state is right-censored at t_J, then we assume that the individual is alive but with unknown state and define

$$C(y_J|y_{J-1},\mathbf{x}) = \sum_{s=1}^{2} P(Y_J = s|Y_{J-1} = y_{J-1},\mathbf{x}).$$

In the summation above, note that $P(Y_J = 1|Y_{J-1} = y_{J-1},\mathbf{x}) = 0$ if $y_{J-1} = 2$, because transitions from state 2 back to state 1 are not possible. If death is observed at t_J, then

$$C(y_J|y_{J-1},\mathbf{x}) = \sum_{s=1}^{2} P(Y_J = s|Y_{J-1} = y_{J-1},\mathbf{x})\, q_{s3}(t_J).$$

So we assume an unknown state at time t_J and then an instant death.

Combining contributions (3.2) for all N individuals, the likelihood function is given by $L = \prod_{i=1}^{N} L_i(\boldsymbol{\theta}|\mathbf{y},\mathbf{x})$. The logarithm of L is maximised over the parameter space to obtain the maximum likelihood estimate (MLE) of $\boldsymbol{\theta}$. Large-sample asymptotic properties of maximum likelihood estimation can be used to estimate the distribution of the MLE.

In case there is an integral in the likelihood, this integral can be approximated numerically. Although many software packages have functions for numerical integration, writing one's own routine might work faster in some cases. The integrals in this chapter are approximated numerically with a user-written implementation of the *composite Simpson's rule*. This implementation uses a uniform grid spacing with an even number of nodes; see Appendix C for details. This approach is also called the *repeated Simpson's rule*.

Implementation-wise, the easiest way to obtain the MLE and to estimate its distribution is to use a general-purpose optimiser. All the main software environments for statistical computing have routines that perform numerical minimisation (or maximisation) of a function of parameters given a set of starting values. General-purpose optimisers for a multi-state survival model will be discussed in more detail in Section 4.5.

For models with piecewise-constant intensities, it is also possible to maximise the likelihood by implementing a scoring algorithm. For a model without right censoring and exact death times, this is shown by Kalbfleisch and Lawless (1985). To take into account the likelihood contributions in (3.2) which include right censoring and death times, the algorithm has to be adapted slightly; details are provided in Section 4.7 and Appendix B. Alternatively, Monte Carlo methods can be used to explore the likelihood. This will be discussed within a Bayesian framework in Chapter 6.

Combining the estimation methods above can save computer time. Starting off with a piecewise-constant approach and the scoring algorithm can provide starting values for more complex models or for Monte Carlo methods.

3.6 Simulation study

This section presents a method to simulate interval-censored three-state survival data. The simulation is illustrated for a progressive process with Gompertz baseline hazards. Extension to other baseline hazard distributions or to extended models is straightforward. In addition, a small simulation study is undertaken to investigate the performance of the piecewise-constant approximation.

For a progressive Gompertz model, define $\beta_{rs} = \log(\lambda_{rs})$ and write $q_{rs}(t) = \exp(\beta_{rs} + \xi_{rs}t)$ for $(r,s) \in \{(1,2),(1,3),(2,3)\}$. Let $T_{rs} = T_{rs|u}$ be the event time for a transition to state s conditional on being in state r at time $u > 0$.

If the state at u is 1, then the time of a transition to the next state can be obtained by taking $T = \min\{T_{12}, T_{13}\}$. If $T = T_{12}$, then the next state is 2. If $T = T_{13}$, the next state is 3. For moving out of state 2, take T_{23}. The transition times conditional on being in a specified state at u can be simulated by sampling from uniform distributions and using the inversion method; see the last part of Section 2.3.3.

After individual trajectories are simulated with known transition times, a study design can be imposed on the trajectories. This will create interval-censored data. As an example, say the time window of the study is 12 years and the observation scheme is bi-yearly. This means that—after the baseline—a snapshot is taken from the simulated trajectories every 2 years.

A small simulation study was conducted to illustrate the piecewise-constant approximation and the effect of study design. The approximation is defined by the individual-specific intervals in the longitudinal data, and differences of this approach with continuous-time estimation are of interest.

In the simulation study, a data set is generated S times, and for each of the data sets the model of interest is estimated. For a model parameter—say θ—this results in estimates $\widehat{\theta}_1, \ldots, \widehat{\theta}_S$. For inference, we compute the mean of the parameter estimates $\overline{\widehat{\theta}} = S^{-1}\sum_{s=1}^{S}\widehat{\theta}_s$, the bias $\overline{\widehat{\theta}} - \theta$, the percentage bias $|100(\overline{\widehat{\theta}} - \theta)\theta^{-1}|$, and the actual coverage percentage (ACP) for 95% confidence intervals. The 95% confidence intervals are estimated as $\widehat{\theta} \pm 1.96 \times SE(\widehat{\theta})$.

The specification of the simulation mimics the study design that is discussed in the application in Section 3.7. The sample size is $N = 200$ and the follow-up time is 12 years. Two study designs will be investigated: Design $\mathcal{A}$, where states are observed yearly, and Design $\mathcal{B}$, where states are observed at baseline and at 4, 8, 9, 10, 11, and 12 years later. In both designs, death times are recorded exactly.

The time scale is age in years. All individuals start in state 1, and baseline age in the simulation ranges uniformly from 1 to 25. This range can refer to transformed age in an application. The parameters for the transition-specific Gompertz distributions are fixed to $\beta_{12} = \beta_{13} = -4$, $\beta_{23} = -3$, $\xi_{12} = \xi_{13} = \xi_{23} = 0.1$.

For the continuous-time estimation, user-written R code was used. The performance of this estimation depends on the approximation of the integral. The composite Simpson's rule is used, where the uniform grid is defined such that time between the nodes is as close as possible to 1/12 year; that is, a month. For the occasional time interval less than a month, four nodes are used. For the piecewise-constant approximation, the R package msm was used, where the piecewise-constant intervals for the transition intensities are defined by the individual-specific time intervals in data.

Table 3.1 presents the results of the simulation study for $S = 200$. For Design $\mathcal{A}$ the results are quite similar for the piecewise-constant approximation and the continuous-time estimation. This is remarkable given that the piecewise-constant approximation uses the time between the observations in the data, which is a year unless the observation is death. For Design $\mathcal{B}$, however, there is a difference in the sense that the piecewise-constant approximation leads to bias inference with respect to the estimation of the λ parameters. This was to be expected given that in Design $\mathcal{B}$ there are 4-year time intervals between some of the observations. Keeping intensities constant for 4 years in the piecewise-constant approximation estimation induces bias. However, Table 3.1 shows that this bias mainly affects the estimation of the

Table 3.1 *Simulation study to compare a piecewise-constant estimation with a continuous-time estimation. Design $\mathcal{A}$ with 12 yearly observations, Design $\mathcal{B}$ with observations at 0, 4, 8, 9, 10, 11, and 12 years. For $S = 200$ replications: mean of estimates, bias, percentage bias, and actual coverage percentage for 95% confidence intervals (ACP). Notation $\lceil x \rceil$ for an absolute value less than x.*

Parameter	Mean		Bias		%Bias		ACP	
	Piecewise-constant estimation							
	$\mathcal{A}$	$\mathcal{B}$	$\mathcal{A}$	$\mathcal{B}$	$\mathcal{A}$	$\mathcal{B}$	$\mathcal{A}$	$\mathcal{B}$
$\beta_{12} = -4$	−3.989	−3.841	0.011	0.159	0.3	4.3	0.96	0.92
$\beta_{13} = -4$	−3.965	−3.826	0.035	0.174	0.9	4.4	0.95	0.89
$\beta_{23} = -3$	−3.024	−2.840	−0.024	0.160	0.8	5.3	0.94	0.90
$\xi_{12} = 0.1$	0.103	0.104	0.003	0.004	3.0	3.8	0.96	0.95
$\xi_{13} = 0.1$	0.100	0.095	$\lceil 0.001 \rceil$	0.005	0.1	5.2	0.97	0.98
$\xi_{23} = 0.1$	0.104	0.100	0.004	$\lceil 0.001 \rceil$	4.1	0.4	0.94	0.91
	Continuous-time estimation							
	$\mathcal{A}$	$\mathcal{B}$	$\mathcal{A}$	$\mathcal{B}$	$\mathcal{A}$	$\mathcal{B}$	$\mathcal{A}$	$\mathcal{B}$
$\beta_{12} = -4$	−4.032	−4.050	−0.032	−0.050	0.8	1.3	0.97	0.94
$\beta_{13} = -4$	−3.987	−3.981	0.013	0.019	0.3	0.5	0.96	0.96
$\beta_{23} = -3$	−3.000	−3.014	$\lceil 0.001 \rceil$	−0.014	$\lceil 0.1 \rceil$	0.5	0.94	0.92
$\xi_{12} = 0.1$	0.103	0.104	0.003	0.004	2.6	4.3	0.96	0.96
$\xi_{13} = 0.1$	0.098	0.097	−0.002	−0.003	1.6	3.5	0.96	0.98
$\xi_{23} = 0.1$	0.101	0.102	0.001	0.002	0.6	1.7	0.94	0.92

λ-parameters—the estimation of the ξ-parameters seems less affected by the wider time intervals in Design $\mathcal{B}$.

The continuous-time estimation performs well for both designs. Even though the choice $S = 200$ is rather small for ACP inference, the tentative results in Table 3.1 show a good coverage.

Estimation by using the piecewise-constant approximation is much faster than using the continuous-time estimation. There is no integral to approximate and, for this three-state model, the transition matrix is available in closed form. The simulation study shows that the piecewise-constant approximation can lead to good results as long as the time between observations is not too long relative to the volatility of the multi-state process.

As already stated in Section 1.2, assuming independence between observation times and a multi-state survival process is often not realistic. There may be settings where the distribution of observation times is covariate-dependent. An example would be the covariate defined by employed versus unemployed. If the employed individuals tend to skip more scheduled observation times than the employed—say due to work-related restrictions —then the approximation bias may induce bias in the estimation of the covariate effect.

It is possible to improve the piecewise-constant approximation by embedding the grid defined by the observation times in a finer grid that is uniform and used for all individual trajectories; see Van den Hout and Matthews (2008). Another option would be to add interval-censored states to the data. For example, if state 1 is observed at t_1 and state 2 is observed at $t_2 = t_1 + 4$, then a record can be added with time $t_1 + 2$ with an unknown state. Van den Hout and Matthews (2010) discuss this approach and show that given certain assumptions estimation can be undertaken by using msm.

3.7 Example

3.7.1 Parkinson's disease study

The methods for the three-state progressive model will be illustrated using a Norwegian study of dementia and survival among patients with Parkinson's disease (Buter et al., 2008). Figure 3.1 depicts the health states measured in the study. Individuals with Parkinson's disease are more likely to develop dementia than individuals without the disease (De Lau et al., 2005) and the onset of dementia is an important predictor of health and need for care. The Parkinson's disease data are kindly provided by Dag Aarsland at the Norwegian Centre for Movement Disorders, Stavanger University Hospital. The data are used here for illustrative purposes only and the results should not be used in isolation to inform clinical practice.

In a region with 220,000 inhabitants in Rogaland County, Norway, an attempt was made to recruit all individuals with recognised idiopathic Parkinson's disease. Parkinson's disease was diagnosed in 245 individuals. There were 18 individuals with clinically significant cognitive impairment, and these were excluded from the study. During follow-up, 3 individuals died before the first assessment, 7 were re-diagnosed as not having Parkinson's disease, and 2 had schizophrenia, leaving $N = 233$ individuals eligible for the present study. Interviews took place in 1993, 1997, 2001, 2002, 2003, 2004, and 2005. There are 3 individuals with a right-censored last state. This data set has also been analysed in a Bayesian framework in Van den Hout and Matthews (2009).

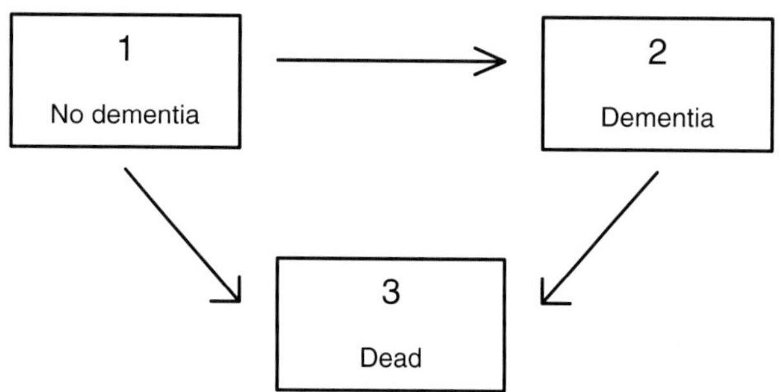

Figure 3.1 *Three-state model for patients with Parkinson's disease.*

Table 3.2 *Frequency table for the Parkinson's disease data: number of times a state was observed (or missing) at baseline and in the follow-up.*

State	Time in years						
	Baseline	4	8	9	10	11	12
1	171	86	42	38	31	29	24
2	62	62	54	39	28	22	19
Dead		79	53	20	18	9	8
Missing		6	5	4	4	3	3

Table 3.2 presents observed frequencies in each state at follow-up times. Because of the progressive nature of dementia, a missing state can sometimes be derived from subsequently observed states. There are 10 individuals with intermediate missing states that cannot be derived in this way. These missing states are noted in the table as the $6+5+4+4+3 = 22$ interval-censored states during follow-up. Interval censoring is not a problem for the estimation of the model, but might bias results if it is substantial. Since there are only 10 persons with missing intermediate interviews and only 2 of them with more than one consecutive interview missing, we assume that the bias is negligible and remove the information on the interval-censored states.

For the data with the interval-censored states removed, Figure 3.2 provides information on the length of time intervals and the number of interviews for men and women together. There are 114 men and 119 women in the data. In total there were 897 observations (total number of interviews, right-censored states, and observed deaths), which means that there were $897 - N = 664$ observed intervals. The number of individuals who died during study time is 187. Note that individuals with only one interview are also included. In the latter case observations consist of the initial state and a time

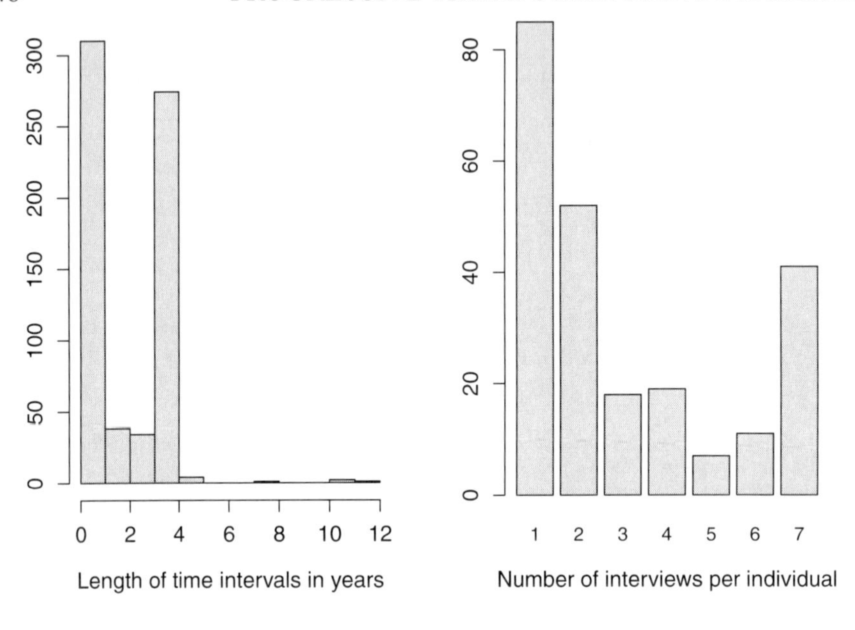

Time intervals		Interviews	
Total number	= 664	Total number	= 707
Mean length	= 2.34	Mean number	= 3.03
Median length	= 1.48	Median number	= 2
Maximum length	= 12	Maximum number	= 7

Figure 3.2 *Length of time intervals and number of interviews for the 233 men and women in the Rogaland Study of Parkinson's Disease.*

of death *or* the initial state and a right-censored state at the end of follow-up. The time interval with maximum length 12 years is the interval of the one individual in the study who was only seen at baseline and was still alive at the end of 2005.

The number of times each pair of states was observed at successive observation times is given in Table 3.3. Note that there are no transitions from state 2 back to state 1, which is in line with the progressive model.

3.7.2 Baseline hazard models

The parametric models for the baseline hazard in Section 3.2 will be fitted to the Parkinson's disease data. The basic model is defined by using exponential hazards for the transition intensities as presented in Section 3.2.1. With three

Table 3.3 *State table for Parkinson's disease data: number of times each pair of states was observed at successive observation times.*

	To state			
	No dementia	*Dementia*	*Dead*	*Right-censored*
From state				
No dementia	250	78	66	3
Dementia	0	146	121	0

possible transitions, this implies three unknown parameters. The model is

$$q_{rs}(t) - \lambda_{rs} = \exp(\beta_{rs}) \qquad \text{for} \qquad (r,s) \in \{(1,2),(1,3),(2,3)\}.$$

The transition-specific λ_{rs} in the exponential model are estimated by $\exp(\beta_{rs})$ to allow maximisation over an unbounded parameter space; that is, $\beta_{rs} \in \mathbb{R}$.

This model is not realistic for the data at hand because it assumes that the hazards do not change over time. Typically, the risk of morbidity and mortality increases with age. However, the exponential is the building block for the extended models and can be seen as the null model or intercept-only model. The model can be fitted in R by using msm, but it is worthwhile to implement this model from scratch to use it as the basis for the implementation of the extended models that cannot be fitted by msm.

Estimated intercept parameters for the Parkinson's disease data are presented in Table 3.4. The model has $-2\log(L_{\max}) = 1377.4$, where $L_{\max}$ is the likelihood value at the maximum likelihood estimate. The estimated intercepts illustrate the increased risk of dying once demented: $\widehat{\lambda}_{13} = \exp(-3.86) = 0.02$ versus $\widehat{\lambda}_{23} = \exp(-1.36) = 0.26$.

For the Gompertz model and the Weibull model, age in the data is transformed. Minimum age at baseline is 36 years. Given an observed age, time t in the models is defined in years as $t = \text{age} - 36 + 1$. The subtraction is used to prevent numerical problems when using the exponential function in the Gompertz model during the maximisation of the likelihood, and the adding of 1 year is to prevent prevent numerical problems with the power function in the Weibull model.

The Gompertz model is given by

$$q_{rs}(t) = \exp(\beta_{rs} + \xi_{rs} t) \qquad \text{for} \quad (r,s) \in \{(1,2),(1,3),(2,3)\}, \tag{3.3}$$

and the Weibull model is given by

$$q_{rs}(t) = \exp(\beta_{rs}) \tau_{rs} t^{(\tau_{rs}-1)} \qquad \text{for} \quad (r,s) \in \{(1,2),(1,3),(2,3)\}. \tag{3.4}$$

Table 3.4 *Estimated parameters for baseline models fitted to the Parkinson's disease data, and information on the maximum value of the log-likelihood function.*

Exponential			
β_{12}	−1.97 (0.11)		
β_{13}	−3.86 (0.38)		
β_{23}	−1.36 (0.08)	$-2\log(L_{\max})$	1377.4

Gompertz		Weibull		Hybrid	
β_{12}	−5.18 (0.65)	β_{12}	−15.25 (2.50)	β_{12}	−14.14 (2.47)
β_{13}	−5.87 (1.65)	β_{13}	−9.01 (5.60)	β_{13}	−9.93 (6.21)
β_{23}	−3.18 (0.64)	β_{23}	−9.85 (2.61)	β_{23}	−3.32 (0.64)
ξ_{12}	0.08 (0.02)	τ_{12}	4.21 (0.63)	τ_{12}	3.93 (0.62)
ξ_{13}	0.05 (0.04)	τ_{13}	2.21 (1.38)	τ_{13}	2.44 (1.54)
ξ_{23}	0.04 (0.01)	τ_{23}	2.94 (0.62)	ξ_{23}	0.04 (0.01)
$-2\log(L_{\max})$					
	1326.2		1324.6		1325.5

For the Weibull model, each τ-parameter is estimated by maximising over $\theta \in \mathbb{R}$, where $\tau = g(\theta) = \exp(\theta)$. The corresponding standard errors are derived by using the delta method; see Section 4.5, or Rice (1995, Section 4.6). Given estimated standard error $\widehat{s}_\theta$ for $\widehat{\theta}$, estimated standard error for $\widehat{\tau}$ is

$$\widehat{s}_\tau = \widehat{s}_\theta \left(g'\left(\widehat{\theta}\right)\right)^2 = \widehat{s}_\theta \left(\exp\left(\widehat{\theta}\right)\right)^2.$$

Combining Gompertz hazard with Weibull hazards within one model is also explored. The best hybrid model for the current data is given by

$$\begin{aligned} q_{12}(t) &= \exp(\beta_{12})\tau_{12}t^{(\tau_{12}-1)} \\ q_{13}(t) &= \exp(\beta_{13})\tau_{13}t^{(\tau_{13}-1)} \\ q_{23}(t) &= \exp(\beta_{23}+\xi_{23}t). \end{aligned}$$

The above parametric baseline models can be compared to a piecewise-constant model. Given that 1 is the minimum of transformed age, a model is fitted with grid points 1, 30, 50, and ∞, which corresponds to ages 36, 66, 86, and ∞, respectively. This piecewise-constant model is given by

$$q_{rs}(t) = \exp\Big(\beta_{rs.0} + \beta_{rs.1} I(30 \le t < 50) + \beta_{rs.2} I(50 < t)\Big),$$

for $(r,s) \in \{(1,2),(1,3),(2,3)\}$. This model has nine parameters and $-2\log(L_{\max}) = 1343.79$. The choice of the grid points is arbitrary up to a certain extent.

The Akaike information criterion (AIC) for model comparison is given by $AIC = -2\log(L_{\max}) + 2k$, where k denotes the number of independent parameters in the model. Taking into account the number of parameters is a penalty for model complexity (Akaike, 1974). The AIC is discussed in more detail in Section 4.7.

From the log-likelihoods in Table 3.4 and the corresponding AICs, it follows that the piecewise-constant model performs better than the exponential model, but worse than the Gompertz and Weibull models, and that the Weibull model outperforms the Gompertz. The hybrid model is in this case true to its name in the sense that its performance is in-between the Weibull and the Gompertz.

Figure 3.3 depicts the fitted transition-specific hazards for the above models. For transition $1 \to 2$ the graph suggests that the subdivision of the age scale in the piecewise-constant model may be chosen more efficient. However, there is lack of data for the older ages and choosing a grid point corresponding to an age above 86 leads to estimation problems. Figure 3.3 also shows that for ages below 86, differences between the models are limited.

The AIC helps us to compare models, but this does not imply that the model with the lowest AIC fits the data well. For checking whether the model is a good description of the data, goodness-of-fit methods are needed. However, the availability of methods for the assessment of goodness of fit is limited for time-dependent multi-state survival models. This will be further discussed in Section 4.9. One of the options is to compare model-based survival with the Kaplan–Meier estimator.

For the current Gompertz and Weibull models, Figure 3.4 compares baseline-state-specific Kaplan–Meier curves with expected survival estimated from the model. In this comparison, the non-parametric Kaplan–Meier estimate is seen as a good representation of the survival in the data and a well-fitting multi-state model should produce expected survival close the Kaplan–Meier curve. This is not a formal test, but the approach should be able to flag structural problems with respect to goodness of fit; see Titman and Sharples (2010a).

Given a fitted three-state survival model for an individual i with transformed age t_i at baseline, a transition matrix can be derived for a given number of years time since baseline. For example, for 5 years since baseline the matrix is $\mathbf{P}(t_i, t_i+5)$ with entries $p_{rs}(t_i, t_i+5)$ for $r,s \in \{1,2,3\}$. If the individual i is in state r at baseline, then $1 - p_{r3}(t_i, t_i+5)$ is the survival probability.

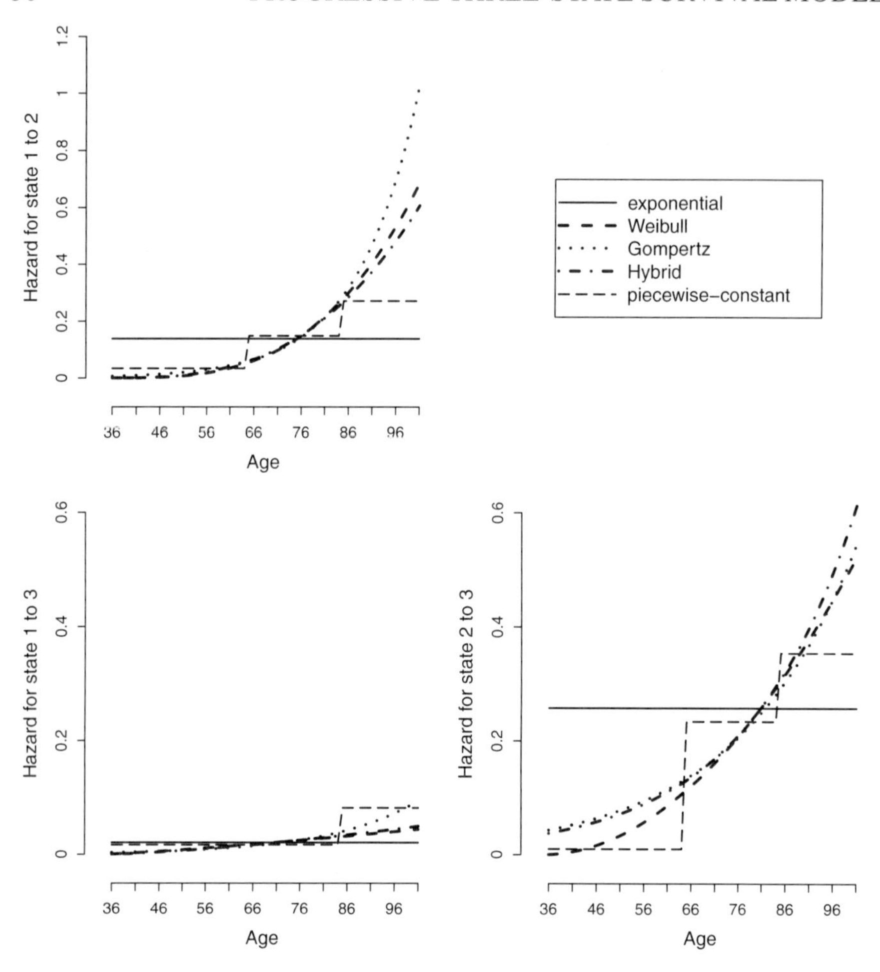

Figure 3.3 *Estimated transition-specific hazards in the baseline hazard models for the Parkinson's disease study. The range of the vertical axes varies.*

Individual expected survival varies because of variation in age at the baseline of the study. The mean trend at 5 years is computed as $1 - \sum_i p_{r3}(t_i, t_i + 5)/N$. Figure 3.4 depicts both the individual probabilities for $r = 1, 2$ and the mean. A similar approach is also used in Kapetanakis et al. (2013).

In terms of goodness of fit, differences in Figure 3.4 between the Gompertz model and Weibull model are minimal. The mean trends are very similar and stay within the 95% band of the Kaplan–Meier estimate. The graphs do not give reason to reject the models.

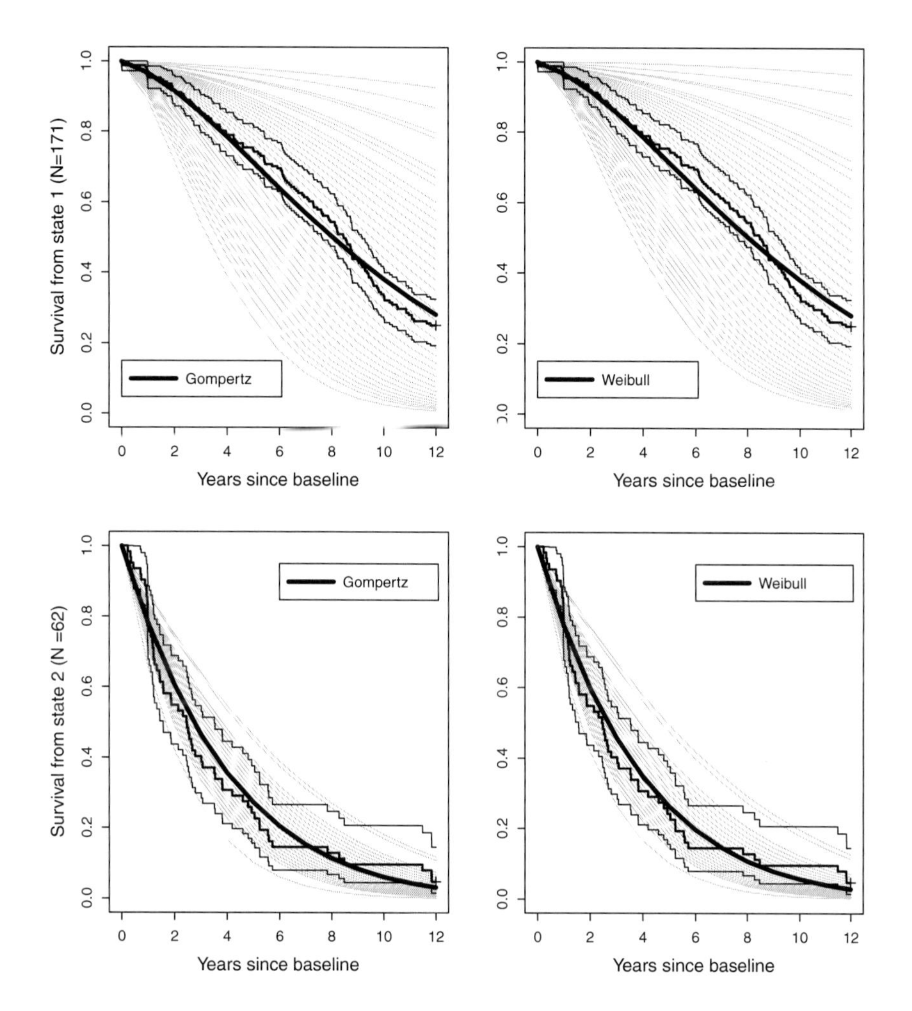

Figure 3.4 *Comparison of Kaplan–Meier curves (with 95% confidence bands) with estimated survival inferred from the Gompertz model and the Weibull model. Smooth black curves for model-based mean trends, grey curves for individual-specific predictions.*

3.7.3 Regression models

The baseline hazard models will be extended by including two covariates in the model, namely, *sex* (0 for men, 1 for women) and *duration* as the number of years the individual had Parkinson's disease before 1993. Minimum, mean, and maximum duration of Parkinson's disease in years is 1, 8.89, and 34,

Table 3.5 *Estimated parameters for the regression models fitted to the Parkinson's disease data. Parameters with an asterisk are restricted to the given value. In addition, information on the maximum value of the log-likelihood function.*

Gompertz		Weibull		
			Unrestricted	*Restricted*
β_{12}	−6.09 (0.69)	β_{12}	−16.36 (2.45)	−16.99 (2.42)
β_{13}	−5.04 (1.90)	β_{13}	−10.14 (7.05)	−4.03 (0.39)
β_{23}	−3.21 (0.67)	β_{23}	−10.13 (2.67)	−9.54 (2.63)
ξ_{12}	0.09 (0.02)	τ_{12}	4.39 (0.62)	4.59 (0.61)
ξ_{13}	0.05 (0.05)	τ_{13}	2.67 (1.75)	1*
ξ_{23}	0.04 (0.01)	τ_{23}	3.05 (0.64)	2.93 (0.64)
$\beta_{12.s}$	−0.56 (0.22)	$\beta_{12.s}$	−0.55 (0.22)	−0.58 (0.20)
$\beta_{13.s}$	−0.19 (0.72)	$\beta_{13.s}$	−0.12 (0.74)	0*
$\beta_{23.s}$	−0.38 (0.17)	$\beta_{23.s}$	−0.37 (0.17)	−0.37 (0.16)
$\beta_{12.d}$	0.08 (0.02)	$\beta_{12.d}$	0.08 (0.02)	0.07 (0.02)
$\beta_{13.d}$	−0.12 (0.10)	$\beta_{13.d}$	−0.11 (0.10)	0*
$\beta_{23.d}$	0.001 (0.02)	$\beta_{23.d}$	0.001 (0.02)	−0.004 (0.02)
$-2\log(L_{\max})$				
	1299.6		1297.7	1301.3

respectively. The log-linear regression models are given by

$$q_{rs}(t) \quad = \quad q_{rs.0}(t)\exp\left(\beta_{rs.s}\, sex + \beta_{rs.d}\, duration\right),$$

for $(r,s) \in \{(1,2),(1,3),(2,3)\}$. Using the Gompertz baseline, we obtain $-2\log(L_{\max}) = 1299.6$, and using the Weibull baseline, we obtain $-2\log(L_{\max}) = 1297.7$. Here we see a continuation of the comparison for the baseline-only models; that is, the Weibull performs better than the Gompertz. Table 3.5 presents the parameter estimates. Note that the estimated regression coefficients are very similar for both choices of the baseline model.

For the Weibull model, the results in Table 3.5 indicate a possible problem: the standard errors for the parameters for the $1 \rightarrow 3$ transition are relatively large. Further modelling shows that it is better to restrict these parameters. The model with restrictions $\tau_{13} = 1$, $\beta_{13.s} = \beta_{13.d} = 0$ has $-2\log(L_{\max}) = 1301.33$ with only nine parameters; see also Table 3.5. In terms of AIC, this second model is much better than the unrestricted Weibull model (AIC = 1319.3 versus AIC = 1321.7, respectively). Notice the change in estimated standard error for β_{13}.

We interpret the results for the Weibull model with the restrictions. As expected, getting older is associated with increased hazard of dementia and death ($\widehat{\tau}_{12}, \widehat{\tau}_{23} > 1$). Longer duration is associated with a higher hazard of dementia ($\widehat{\beta}_{12.d} > 0$), and women ($sex = 1$) with dementia have a lower hazard of death than men ($\widehat{\beta}_{23.s} < 0$). With respect to the onset of dementia, the hazard ratio for women versus men is $\exp(\widehat{\beta}_{12.s}) = 0.58$.

Although the estimated coefficients for the log-linear hazard model are informative when, for instance, comparing men and women, they do not provide information with regard to prediction. Estimated transition probabilities are more suitable for this purpose. For example, for a woman aged 65 with a duration equal to 5 years, the estimated transition probability matrix for the next 5 years is given by

$$\widehat{\mathbf{P}}(t_1 = 30, t_2 = 35 | sex = 1, duration = 5) = \begin{pmatrix} 0.75 & 0.13 & 0.12 \\ 0 & 0.56 & 0.44 \\ 0 & 0 & 1 \end{pmatrix},$$

where $t_1 = 30$ and $t_2 = 35$ correspond to ages 65 and 70. This means, for instance, that if the woman is demented at age 65 and has had Parkinson's disease for 5 years, then the probability that she dies in the next 5 years is 0.44. For a man of the same age and with the same duration, the matrix is

$$\widehat{\mathbf{P}}(t_1 = 30, t_2 = 35 | sex = 0, duration = 5) = \begin{pmatrix} 0.64 & 0.19 & 0.17 \\ 0 & 0.43 & 0.57 \\ 0 & 0 & 1 \end{pmatrix}.$$

In practice, of interest will often be the probability to be demented at a certain age, conditional on still being alive at that age. If the woman above is not demented at 65, then this probability is $P(\text{demented}|\text{not dead}) = P(\text{demented})/(1 - P(\text{dead}))$, which is estimated at $0.13/(1 - 0.12) = 0.15$. For the man, the estimate is 0.23.

Note that the probability $P(\text{demented}|\text{not dead})$ is a complex function of model parameters. Therefore, propagating the uncertainty in the estimation of the model parameters into the uncertainty of the estimate for $P(\text{demented}|\text{not dead})$ is not a straightforward application of the delta method. Simulation and the asymptotic behaviour of maximum likelihood estimates can be of help here. The following simulation scheme can be used to derive confidence intervals for any statistic that is a function of the parameters of the multi-state survival model; for more details see Section 4.5.

Given a fitted model, consider the multivariate normal distribution with expectation equal to the maximum likelihood estimate of the parameter vector and the covariance matrix equal to the estimated covariance matrix at the optimum. The sample variation in the estimation of $P(\text{demented}|\text{not dead})$ is evaluated by drawing parameter values from this multivariate distribution and computing $P(\text{demented}|\text{not dead})$ for each of the drawn values.

Using 1000 simulated parameter vectors, we obtain for $P(\text{demented}|\text{not dead})$ the 95% confidence interval (0.09, 0.32) for women, and interval (0.12, 0.45) for men. These overlapping intervals show that for this specific conditional probability there is no clear statistical difference between men and women.

Chapter 4

General Multi-State Survival Model

4.1 Discrete-time Markov process

A *stochastic process* is a collection of random variables $\{Y_t | \, t \in \mathcal{U}\}$, where $\mathcal{U}$ is some index set. If $\mathcal{U} = \{0, 1, 2, \ldots\}$, then the process is in discrete time. If $\mathcal{U} = (0, \infty)$, then the process is in continuous time. Variable Y_t is the *state* of the process at time t. The set $\mathcal{S}$ of all possible values of Y_t is the *state space*. Here we limit ourselves to $\mathcal{S}$ being the finite discrete-state space $\{1, 2..., D\}$ for some integer D. The theory for basic discrete-time Markov processes is well established. The material in this section can also be found in Norris (1997), Kulkarni (2011), or in any other textbook on stochastic processes.

Denote discrete-time processes by $\{Y_n | \, n = 0, 1, 2, \ldots\}$. For a sequence of discrete random variables $Y_0, Y_1, \ldots, Y_J$ we can write the joint probability as a product of conditional probabilities:

$$\begin{aligned} P(Y_0 = y_0, Y_1 = y_1, Y_2 = y_2, \ldots, Y_J = y_J) \\ = \; & P(Y_0 = y_0) P(Y_1 = y_1 | Y_0 = y_0) P(Y_2 = y_2 | Y_1 = y_1, Y_0 = y_0) \\ & \times \cdots \times P(Y_J = y_J | Y_{J-1} = y_{J-1}, \ldots, Y_0 = y_0). \end{aligned} \tag{4.1}$$

A discrete-time stochastic process is called a *Markov chain* if the conditional distribution of any future state, given the present state, is unaffected by any knowledge of the history of the process. For the above sequence $Y_0, Y_1, \ldots, Y_J$, this would imply that (4.1) simplifies to

$$\begin{aligned} P(Y_0 = y_0, Y_1 = y_1, Y_2 = y_2, \ldots, Y_J = y_J) \\ = P(Y_0 = y_0) P(Y_1 = y_1 | Y_0 = y_0) P(Y_2 = y_2 | Y_1 = y_1) \\ \times \cdots \times P(Y_J = y_J | Y_{J-1} = y_{J-1}). \end{aligned}$$

A discrete-time Markov chain is *time homogeneous* if the conditional probability $P(Y_{n+1} = s | Y_n = r)$ does not depend on n. It that case, define

$\bar{p}_{rs} = P(Y_{n+1} = s|Y_n = r)$ as the *transition probability* from state r to state s. The bar-notation is used to distinguish this probability from the probabilities that will be defined for the continuous-time model. The $D \times D$ matrix $\bar{\mathbf{P}}$ with $\bar{p}_{rs}$ as the (r,s) entry, for all $r,s \in \mathcal{S}$, is the *transition matrix* of the chain. It follows that the rows of $\bar{\mathbf{P}}$ sum up to 1.

Conditional probability $\bar{p}_{rs}^{(n)} = P(Y_n = r|Y_0 = s)$ is called the *n-step transition probability*. The *n-step transition probability matrix* is $\bar{\mathbf{P}}^{(n)} = \bar{\mathbf{P}}^n$. The Chapman–Kolmogorov equation for transition probabilities is given in matrix form by $\bar{\mathbf{P}}^{(n+m)} = \bar{\mathbf{P}}^n\bar{\mathbf{P}}^m$. When $\bar{\mathbf{P}}$ is given, long-term behaviour of the process can be derived easily. For example, the time spent in current state r until the next transition has a geometric distribution; that is,

$$P(\text{time spent in state } r > n|\text{current state is } r) = (1 - \bar{p}_{rr})^n.$$

This book's main topic is continuous-time survival processes. The discrete-time processes are mentioned here for two reasons: first to have a clear distinction—if actual time is involved, transitions in a discrete-time process are timed on a uniform grid where the time between the grid points is not part of the model. Because of this, one can speak of steps of the process since the time intervals are the same throughout. For each time interval, at most one transition is possible. In contrast, transitions in a continuous-time process are timed on a continuum.

The second reason to pay attention to discrete-time processes is that these processes can be used to approximate continuous-time processes; see, for example, the applications in Lièvre et al. (2003) and Cai et al. (2010). There is also work on multi-state processes where both time formulations are used within one model. For a three-state progressive illness-death model, Dinse (1988) uses discrete time for death and continuous time for the onset of the illness. Chapter 8 discusses statistical models for discrete-time Markov processes.

4.2 Continuous-time Markov processes

A continuous-time stochastic process $\{Y_t|t \in (0,\infty)\}$ is a *Markov chain* on state space $\mathcal{S}$ if

$$P\big(Y_{u+t} = s\big|Y_u = r,\ \{Y_v|0 \le v < u\}\ \big) = P\big(Y_{u+t} = s|Y_u = r\big),$$

for all $r,s \in S$ and $u,t \ge 0$. The Markov chain is *time homogeneous* if, for $t,u \ge 0$,

$$P(Y_{u+t} = s|Y_u = r) = P(Y_t = s|Y_0 = r).$$

These are the standard definitions; the above phrasing is similar to the text in Kulkarni (2011). Other references for basic theory on continuous-time processes are, for example, Cox and Miller (1965) and Norris (1997). In addition, the publication Kalbfleisch and Lawless (1985) is important because it discusses the analysis of longitudinal data by combining the theory on continuous-time Markov models with maximum likelihood estimation.

If we define for a time-homogeneous chain *transition probabilities*

$$p_{rs}(t) = P\big(Y_t = s | Y_0 = r\big),$$

then the matrix $\mathbf{P}(t)$ containing these probabilities such that r denotes the row and s the column is called the *transition probability matrix*. The rows of $\mathbf{P}(t)$ sum up to 1. The Chapman–Kolmogorov equations for a continuous-time Markov chain are $\mathbf{P}(u+t) = \mathbf{P}(u)\mathbf{P}(t)$ for any $u, t > 0$.

Note that transition probabilities for a continuous-time Markov chain are functions of time t. These functions can be derived from the *transition rates* for the chain. For a time-homogeneous chain, the *transition rates* are given by

$$q_{rs} = \lim_{\Delta \downarrow 0} \frac{P\big(Y_{t+\Delta} = s | Y_t = r\big)}{\Delta}, \tag{4.2}$$

for $r, s \in \mathcal{S}, r \neq s$, and any $t \geq 0$. The matrix with off-diagonal (r,s)-entries q_{rs} and diagonal entries $q_{rr} = -\sum_{s \neq r} q_{rs}$ is called the *generator matrix* $\mathbf{Q}$. Per definition, the rows of $\mathbf{Q}$ sum to zero. The transition rates for a continuous-time Markov chain are also called *transition intensities* or *transition hazards*.

The *Kolmogorov differential equations* state the direct link between the generator matrix and the transition probability matrix for given time t. The forward equations and backward equations are

$$\mathbf{P}'(t) = \mathbf{P}(t)\mathbf{Q} \qquad \text{and} \qquad \mathbf{P}'(t) = \mathbf{Q}\mathbf{P}(t),$$

respectively. Given the generator matrix, the solution for $\mathbf{P}(t)$ subject to $\mathbf{P}(0) = \mathbf{I}$ is the matrix exponential

$$\mathbf{P}(t) = \sum_{m=0}^{\infty} \frac{(t\mathbf{Q})^m}{m!} = \exp(t\mathbf{Q}). \tag{4.3}$$

In Chapter 3, the progressive three-state survival model is presented by extending the approach for a univariate survival process to a process with more than one event. This was possible because of the progressive nature of the three-state process. Given interval censoring with state 1 observed at t_1 and state 2 at t_2, it follows that there was exactly one event in $(t_1, t_2]$. For the case with state 1 observed at t_1 and death t_2, it follows that there are only two possible trajectories underlying the observations: either $1 \to 3$ with exactly one event, or $1 \to 2 \to 3$ with exactly two events.

The general multi-state survival model allows the definition of reversible processes; for example, a three-state survival model where back-and-forth transitions between states 1 and 2 are possible. And it is because there are no restrictions on the number of transitions between living states, that differential equations come into play. For a rigorous explanation of the forward equations and backward equations, see the aforementioned literature. Here we provide a heuristic explanation.

First note that with (4.2), we have for small $h > 0$

$$p_{rs}(h) \approx hq_{rs} \qquad \text{for} \quad r \neq s. \tag{4.4}$$

Using (4.4) and $p_{rr}(h) = 1 - \sum_{s \neq r} p_{rs}(h)$, it follows that

$$p_{rr}(h) \approx 1 + hq_{rr}\,. \tag{4.5}$$

With the Chapman–Kolmogorov equations we have

$$p_{rs}(t+h) = \sum_{k \in \mathcal{S}} p_{rk}(t) p_{ks}(h),$$

and using (4.4) and (4.5), we obtain

$$\begin{aligned} p_{rs}(t+h) &\approx (1 + hq_{ss}) p_{rs}(t) + \sum_{k \in \mathcal{S}, k \neq s} p_{rk}(t) h q_{ks} \\ &\approx p_{rs}(t) + h \sum_{k \in \mathcal{S}} p_{rk}(t) q_{ks}. \end{aligned}$$

This can be written as

$$\frac{p_{rs}(t+h) - p_{rs}(t)}{h} \approx \sum_{k \in \mathcal{S}} p_{rk}(t) q_{ks},$$

which leads to the forward equation by taking $h \to 0$.

The focus of the current book is on multi-state survival models which are defined as continuous-time multi-state processes with exactly one absorbing state. The row in the generator matrix $\mathbf{Q}$ that corresponds to the absorbing state has all entries equal to zero. This is in line with the interpretation of the off-diagonal entries q_{rs}, $r \neq s$, as rates in (4.2). Once the process is in an absorbing state r, the rate of moving to another state s is zero. As a consequence, the diagonal entry q_{rr} is also zero.

The row in the transition probability matrix $\mathbf{P}(t)$ that corresponds to an absorbing state has the diagonal entry equal to one and all off-diagonal entries equal to zero. This also makes sense: given that the process is in an absorbing state, the probability to be in that state at any future time is one. This property can also be derived from (4.3), which implies for that for $t \downarrow 0$ we have

$$\mathbf{P}(t) = \mathbf{I} + t\mathbf{Q} + \mathcal{O}(t^2).$$

Notation $f(t) = \mathcal{O}(t^2)$ means that in the limit $t \to 0$, $f(t)/t^2 \leq C$ for some finite C. Hence, if all entries in rth row in $\mathbf{Q}$ are zero, then the (r, r) entry in $\mathbf{P}(t)$ is equal to 1.

For some multi-state processes, $\mathbf{P}(t) = \exp(t\mathbf{Q})$ is available in closed form; see, for example, the progressive model in Chapter 3. In general, however, the computation of the exponential of a square matrix is not straightforward; see Moler and Van Loan (2003) for a review that covers many methods including matrix-decomposition approaches and the Padé approximation.

For a $D \times D$ matrix $\mathbf{U}$ with D linearly independent eigenvectors, eigenvalue decomposition can be used for the matrix exponential. It follows that $\exp(\mathbf{U}) = \mathbf{A}\exp(\mathbf{B})\mathbf{A}^{-1}$, where $\exp(\mathbf{B})$ is the diagonal matrix with the exponentiated eigenvalues of $\mathbf{U}$ and $\mathbf{A}$ is the matrix with the eigenvectors of $\mathbf{U}$ as the columns. In case the eigenvectors are not linearly independent, the exponential can be approximated by using a finite summation in (4.3), but this is in general not recommended (Moler and Van Loan, 2003).

To illustrate the efficiency of the computations that are involved in solving the differential equations, consider a four-state survival model with intensity matrix given by

$$\mathbf{Q} = \begin{pmatrix} q_{11} & q_{12} & q_{13} & q_{14} \\ q_{21} & q_{22} & q_{23} & q_{24} \\ q_{31} & q_{32} & q_{33} & q_{34} \\ q_{41} & q_{42} & q_{43} & q_{44} \end{pmatrix} = \begin{pmatrix} -0.2 & 0.1 & 0 & 0.1 \\ 0.2 & -0.6 & 0.2 & 0.2 \\ 0 & 0 & -0.3 & 0.3 \\ 0 & 0 & 0 & 0 \end{pmatrix}.$$

Let $\tilde{\mathbf{P}}_M(t) = \sum_{m=0}^{M}(t\mathbf{Q})^m/m!$. Comparing this approximation to the matrix obtained by using the eigenvalue decomposition, for $M = 25$, there are numerical differences from the 6th decimal place onward. For $M = 30$, differences start at the 9th decimal place. When using software R for the computations, there are no discernible differences in central-processing-unit (CPU) time for using either the eigenvalue decompositions or the finite summation for both choices of M.

Kulkarni (2011) presents an algorithm to compute the transition matrix $\mathbf{P}(t)$ within a given numerical accuracy. For a given $\varepsilon > 0$, the *uniformisation algorithm* iteratively computes $\tilde{\mathbf{P}}_M(t)$ using the smallest M needed such that $|\tilde{p}_{rs}(t) - p_{rs}(t)| < \varepsilon$, for all r, s. In the above example, specifying $\varepsilon = 1 \times 10^{-6}$ requires $M = 20$.

More information on closed-form expressions for $\mathbf{P}(t)$ and computational aspects is given in Appendix A.

4.3 Hazard regression models for transition intensities

Transition-specific hazard regression models can be defined for those $r, s \in S$ between which a transition is possible according to the specified multi-state process. The model that will be used throughout this book combines a baseline hazard with log-linear regression and is given by

$$q_{rs}(t) = q_{rs}(t|\mathbf{x}(t)) = q_{rs.0}(t) \exp\left(\boldsymbol{\beta}_{rs}^\top \mathbf{x}(t)\right), \tag{4.6}$$

for parameter vector $\boldsymbol{\beta}_{rs} = (\beta_{rs.1}, \beta_{rs.2}, ..., \beta_{rs.p})^\top$ and covariate vector $\mathbf{x}(t) = \left(x_1(t), x_2(t), ..., x_p(t)\right)^\top$, which has no intercept. Baseline hazard in (4.6) is given by $q_{rs.0}(t)$. Examples of parametric baseline hazards are

$$\begin{array}{llll}
\text{exponential:} & q_{rs.0}(t) & = \lambda_{rs} & \lambda_{rs} > 0 \\
\text{Weibull:} & q_{rs.0}(t) & = \lambda_{rs}\tau_{rs}t^{\tau_{rs}-1} & \lambda_{rs}, \tau_{rs} > 0 \\
\text{Gompertz:} & q_{rs.0}(t) & = \lambda_{rs}\exp(\xi_{rs}t) & \lambda_{rs} > 0 \\
\text{log-logistic:} & q_{rs.0}(t) & = \frac{\lambda_{rs}\rho_{rs}(\lambda_{rs}t)^{\rho_{rs}-1}}{1+(\lambda_{rs}t)^{\rho_{rs}}} & \lambda_{rs}, \rho_{rs} > 0.
\end{array}$$

The general definition of the log-linear model (4.6) shows that many contemporary statistical regression methods can be incorporated in the analysis of multi-state data. The simplest multi-state model is the intercept-only time-homogeneous model given by $q_{rs}(t) = q_{rs} = \exp(\beta_{rs})$, in which case there are as many parameters to estimate as the number of transitions $r \to s$ that are possible according to the specified process. An example of a more complex model is a frailty model, where for some of the transitions $r \to s$ there is a random cluster effect, for example, $q_{rs}(cluster\ j) = \exp(\beta_{rs.0} + \beta_{rs.j})$ with the assumption that $\beta_{rs.j} \sim N(0, \sigma^2)$ for unknown σ. Frailty models are discussed in detail in Chapter 5.

Baseline hazard definitions can vary across transitions. An example is specifying Weibull hazards for transition between living states, and Gompertz hazards for transitions into the dead state.

In practice, there may be limited information in the data on certain transitions. Typically, one starts with exponential hazards and tries to extend the model step by step by adding time dependency or covariates. Restrictions on parameters can be of help in this forward selection of models. Restrictions can take two forms. Either parameters are restricted to be zero, or parameters are restricted to be the same. An example of the former is a restriction on regression coefficients for a transition for which data are sparse. An example of the latter is to restrict the ξ-parameters to be the same in the Gompertz baseline hazards for all transitions into the dead state. That is, $\xi_{rD} \stackrel{\text{d}}{=} \xi_D$ for D the dead state and for those r for which $r \to D$ is modelled.

The definition of the log-linear regression hazard model (4.6) allows for a time-dependent covariate vector $\mathbf{x}(t)$. Following Kalbfleisch and Prentice (2002, Section 6.3) we distinguish *internal covariates* and *external covariates*.

In the context of survival analysis, an internal time-dependent covariate is a variable that can only be measured while the individual is alive. Examples are blood pressure, lung function, and being married. In contrast, values of external time-dependent covariates are not affected by the individuals under study, or conditional on survival.

An external covariate is *fixed* when it can be measured in advance and does not change during the study, and *defined* when time-dependent values are known up to any time for each individual in the study. Examples of the former are covariates fixed at baseline values, and the archetype example of the latter is age during the follow-up. Note that age can be used both as a fixed external covariate, that is, baseline age in a clinical trial, and as a defined external covariate when used as a time-dependent covariate in a hazard regression model. An external time-dependent covariate can also be *stochastic*; for example, levels of air pollution.

During follow-up, internal covariate values are the output of a stochastic process. In the presence of censoring, this process is only partly observed. With respect to the estimation of the model parameters, the stochastic process can be taken into account by a piecewise-constant approximation of the hazard; see the next section. Using such an approximation means that the distinction interval versus external does not play a role in the estimation. Prediction of the multi-state process in the presence of a time-dependent internal covariate is, however, inherently more difficult as it requires a prediction of the stochastic process of the covariate.

4.4 Piecewise-constant hazards

The introduction to continuous-time Markov processes in Section 4.2 is for processes with constant generator matrix $\mathbf{Q}$. Although it is relatively straightforward to define time-dependent models for the entries $\mathbf{Q}$, as shown in Section 4.3, it is not straightforward to deal with this time dependency in inference. The problem is that the time dependency induces the Kolmogorov forward equations to be time-dependent, which makes the matrix-exponential solution in (4.3) not applicable.

It is possible to define a time-dependent model and circumvent the time-dependent differential equations. Hubbard et al. (2008) show how the time scale of a time-dependent Markov process can be transformed to a time scale on which the process is time-independent. They present a method which jointly estimates the parameters for the time transformation and the parameters for the time-independent Markov process. Another option is to solve the time-dependent differential equations numerically at the level of the equations; see Titman (2011) and the reference therein. Both options are promising, and will no doubt be investigated and applied in the near future.

In this book, we use a more simple method which consists of dealing with the time dependency piecewise-constantly and using the matrix-exponential solution iteratively. As with all methods for approximation, it worthwhile to investigate the performance of the method by varying specifications. An example of this is given in Section 3.6 were a piecewise-constant method was investigated in a simulation study.

For the univariate survival model in Chapter 2 and the three-state progressive survival model in Chapter 3, we distinguished *piecewise-constant hazard models* from *piecewise-constant approximations* of parametric shapes. For the general multi-state survival models in the remainder of this book, the *piecewise-constant approximation* will be the default approach for handling time-dependent hazards.

Consider the time interval $(t_1, t_2]$ with observed states at t_1 and at t_2. If we work with constant hazards within this time interval, the generator matrix $\mathbf{Q}$ is defined for time t_1 using regression model (4.6), and the transition probability matrix $\mathbf{P}(t_2 - t_1)$ is subsequently defined for elapsed time $t_2 - t_1$ using $\mathbf{Q}$. Notation $\mathbf{Q}(t_1)$ and $\mathbf{P}(t_1, t_2)$ will be used from now on to stress the dependence on t_1.

Variations of the above are possible. For example, for $(t_1, t_2]$, matrix $\mathbf{Q}$ can also be defined midway through the time interval; that is, for time $(t_1 + t_2)/2$. This option, however, induces the need for further approximation in case values of an internal time-dependent covariate are only available at times when a living state is observed. If the state at t_2 is the dead state, then one has to decide how to approximate the value of the covariate midway through the time interval.

In longitudinal data, trajectories consist of repeated observations of the state. Say states are observed at times t_1, t_2, and t_3. In that case, the transition matrix for $(t_1, t_3]$ is given by $\mathbf{P}(t_1, t_3) = \mathbf{P}(t_1, t_2)\mathbf{P}(t_2, t_3)$, where both matrices at the right-hand side are derived using constant hazards. If time between t_2 and t_3 is long relative to the volatility of the multi-state process, the piecewise-constant approximation can be improved by imputing an addition time point u and using $\mathbf{P}(t_1, t_3) = \mathbf{P}(t_1, t_2)\mathbf{P}(t_2, u)\mathbf{P}(u, t_3)$. This only works, of course, when $\mathbf{Q}(u)$ can be computed.

The grid for the piecewise-constant hazards can be imposed or can be defined by the data at hand. The former consists of defining a grid and embedding observed time intervals into this grid. For example, say the grid is defined by $u_1, ..., u_M$. For observed time interval $(t_1, t_2]$, determine j_1 and j_2 such that $u_{j_1} < t_1 \leq u_{j_1+1}$ and $u_{j_2} < t_2 \leq u_{j_2+1}$. The transition probability matrix for $(t_1, t_2]$ is then defined by

$$\mathbf{P}(t_1, t_2) = \\ \mathbf{P}\left(t_1, u_{j_1+1} | \mathbf{Q}(u_{j_1})\right) \mathbf{P}\left(u_{j_1+1}, u_{j_1+2} | \mathbf{Q}(u_{j_1+1})\right) \times \cdots \times \mathbf{P}\left(u_{j_2}, t_2 | \mathbf{Q}(u_{j_2})\right).$$

For this approach covariate values $\mathbf{x}(t)$ are needed at all grid points $u_1, ..., u_M$. For an internal time-dependent covariate, these values may not be available.

Alternatively, the grid for the piecewise-constant hazards is defined by the data at hand. In that case, the transition matrix for observed interval $(t_1, t_2]$ is $\mathbf{P}(t_1, t_2)$, which is derived using $\mathbf{Q}(t_1)$, and the piecewise-constant approach extends to piecewise-constant time dependency of time-dependent covariates.

4.5 Maximum likelihood estimation

Estimation of model parameters is undertaken by maximising the log-likelihood. Because of the interval censoring, the likelihood is constructed using transition probabilities. The same expressions can be used as in Chapter 3, but now they refer to any possible number of states (including exactly one absorbing dead state). For the transition into the dead state, we assume that the time is known.

Let the dead state be denoted by integer D and the living states be numbered by $1, 2, .., D-1$. For individual i there are observation times $t_1, ..., t_J$, where the state at t_J is allowed to be right-censored. The number of observations J is allowed to vary across individuals. Let $\mathbf{y}$ denote the observed trajectory defined by the series of states $y_1, ..., y_J$ corresponding to the times $t_1, ..., t_J$. Using the Markov assumption, the contribution of the individual to the likelihood conditional on the first state is given by

$$
\begin{aligned}
L_i(\boldsymbol{\theta}|\mathbf{y},\mathbf{x}) &= P(Y_J = y_J, \ldots, Y_2 = y_2 | Y_1 = y_1, \boldsymbol{\theta}, \mathbf{x}) \\
&= \left(\prod_{j=2}^{J-1} P(Y_j = y_j | Y_{j-1} = y_{j-1}, \boldsymbol{\theta}, \mathbf{x}) \right) C(y_J | y_{J-1}, \boldsymbol{\theta}, \mathbf{x}), \quad (4.7)
\end{aligned}
$$

where $\boldsymbol{\theta}$ is the vector with all the model parameters, and $\mathbf{x}$ denotes the vector with the covariate values.

If a living state at t_J is observed, then

$$C(y_J | y_{J-1}, \mathbf{x}) = P(Y_J = y_J | Y_{J-1} = y_{J-1}, \boldsymbol{\theta}, \mathbf{x}).$$

If the state is right-censored at t_J, then

$$C(y_J | y_{J-1}, \mathbf{x}) = \sum_{s=1}^{D-1} P(Y_J = s | Y_{J-1} = y_{J-1}, \boldsymbol{\theta}, \mathbf{x}).$$

If death is observed at t_J, then

$$C(y_J | y_{J-1}, \mathbf{x}) = \sum_{s=1}^{D-1} P(Y_J = s | Y_{J-1} = y_{J-1}, \boldsymbol{\theta}, \mathbf{x}) \, q_{sD}(t_{J-1} | \boldsymbol{\theta}, \mathbf{x}).$$

Combining contributions (4.7) for all N individuals, the likelihood function is given by $L = \prod_{i=1}^{N} L_i(\boldsymbol{\theta}|\mathbf{y},\mathbf{x})$. Maximising the likelihood function over the parameter space for $\boldsymbol{\theta}$ can be undertaken by using a general-purpose optimiser, a scoring algorithm, Monte Carlo methods, or by a combination of these methods.

The scoring algorithm will be discussed in the next section.

A general-purpose optimiser was used by Kay (1986), Davison and Ramesh (1996), and Satten and Longini (1996), where the latter implemented a downhill simplex method. Marshall et al. (1995) programmed an optimiser explicitly for multi-state survival models with interval-censored transitions among the living states, and known death times. This optimiser uses the likelihood function and either finite differences to obtain numerical approximations of the derivatives, or explicit expressions for the first derivatives; see also Marshall and Jones (1995).

Alioum and Commenges (2001) extend Marshall et al. (1995) by including time dependency using piecewise-constant intensities. The cut-points for the change in time dependency are handled by introducing additional covariates. The resulting model is a piecewise-constant hazard model as defined in Section 4.4. Commenges (2002) suggests to use the approach in Alioum and

Commenges (2001) to deal with parametric time dependency, but does not use this in applications.

Most of the software for statistical computing have general-purpose optimisers as standard functions. In R, the free software environment for statistical computing and graphics (R Core Team, 2013), `optim` is such a function. It can be used to maximise a likelihood function without providing explicit expressions for the derivatives. The default method for optimisation is due to Nelder and Mead (1965). This method is relatively slow but robust in the sense that it is good at dealing with irregular functions and rapidly changing curvature. Another choice in `optim` is to use the method that is described in Broyden (1970), Fletcher (1970), Goldfarb (1970), and Shanno (1970). This method is called the BFGS method and it is faster than Nelder–Mead but less robust.

For both optimisation methods, `optim` will return the numerically differentiated Hessian matrix if requested. If the log-likelihood function is maximised, the Hessian can be used to derive the estimated covariance matrix for the maximum likelihood estimate.

As with all general-purpose optimisers, it is recommended to investigate the maximisation by varying starting values, exploring different methods for optimisation, taking note of convergence assessment, and inspecting the estimated variance-covariance matrix.

Covariance of a function of estimated model parameters can be derived by using the multivariate delta method, or by using simulation. An important example of such a function is the matrix with the transition probabilities for a specified time interval.

The multivariate delta method is a generalisation of the univariate method as used in Section 3.7.2. Consider estimator $\widehat{\boldsymbol{\theta}}_n$ which depends on sample size n, and the vector $\boldsymbol{\theta}_0$ with the true values. Assume that

$$\sqrt{n}\left(\widehat{\boldsymbol{\theta}}_n - \boldsymbol{\theta}_0\right) \xrightarrow{D} N(\mathbf{0}, \boldsymbol{\Sigma}) \qquad \text{for} \qquad n \to \infty,$$

where the left-hand-side arrow denotes convergence in distribution, and $\boldsymbol{\Sigma}$ is a variance-covariance matrix. When a function g is differentiable at $\boldsymbol{\theta}_0$, then if follows that

$$\sqrt{n}\big(g(\widehat{\boldsymbol{\theta}}_n) - g(\boldsymbol{\theta}_0)\big) \xrightarrow{D} N(\mathbf{0}, \mathbf{D}^\top \boldsymbol{\Sigma}\mathbf{D}),$$

where $\mathbf{D}$ is the gradient of g at $\boldsymbol{\theta}_0$. The proof follows from a Taylor series expansion for $g(\widehat{\boldsymbol{\theta}}_n)$ in a neighbourhood of $\boldsymbol{\theta}_0$; see Casella and Berger (2002, Section 5.5.4) for more details, or Agresti (2002, Chapter 14) for a summary. So for large n, $g(\widehat{\boldsymbol{\theta}}_n)$ has a distribution similar to the normal with mean $g(\boldsymbol{\theta}_0)$ and variance $\mathbf{D}^\top \boldsymbol{\Sigma}\mathbf{D}/n$.

Applying the delta method to the maximum likelihood for the multi-state model parameters, given estimated covariance matrix $\widehat{\mathbf{V}}_{\boldsymbol{\theta}}$ for maximum likelihood estimate $\widehat{\boldsymbol{\theta}}$, estimated covariance for $g(\widehat{\boldsymbol{\theta}})$ is given by

$$\widehat{\mathbf{V}}_{g(\boldsymbol{\theta})} = \left(\frac{\partial g}{\partial \boldsymbol{\theta}}\right)^{\top} \widehat{\mathbf{V}}_{\boldsymbol{\theta}} \left(\frac{\partial g}{\partial \boldsymbol{\theta}}\right). \tag{4.8}$$

If the scoring algorithm is used to maximise the likelihood and the function g of interest is the matrix with the transition probabilities, then the first-order derivative in (4.8) is available as it is used in the algorithm; see, for example, Gentleman et al. (1994). If the function of interest is a matrix with transition probabilities and the piecewise-constant approximation is used for parametric time dependency, then the derivative is a derivative of a multiplication of transition matrices. This will be illustrated in Section 4.8.2.

An alternative to the delta method is to use simulation. In that case, a parameter vector $\boldsymbol{\theta}^{(b)}$ is draw from $N(\widehat{\boldsymbol{\theta}}, \widehat{\mathbf{V}}_{\boldsymbol{\theta}})$, for $b = 1, ..., B$, and summary statistics such as mean and covariance are derived from $g(\boldsymbol{\theta}^{(1)}), ..., g(\boldsymbol{\theta}^{(B)})$; see Mandel (2013) for a justification of this method, and, for example, Aalen et al. (1997) for an application.

If a general-purpose optimiser is used and a numerically derived $\widehat{\mathbf{V}}_{\boldsymbol{\theta}}$ is part of the output of the optimiser, then using the simulation method can serve as the default choice. A motivation for choosing the simulation as the default is that this method takes into account potential boundary solutions. For example, if one of the transition probabilities is close to zero, the corresponding 2.5% and 97.5% quantiles of the simulated probabilities can be used to estimate the skewed 95% confidence interval. Another advantage is that the simulation method is easier to implement if a piecewise-constant approximation is used.

When Monte Carlo methods are used for Bayesian inference, sampled $\boldsymbol{\theta}^{(b)}$, for $b = 1, ..., B$, are available from the posterior directly; see Chapter 5. Summary statistics can then be derived from $g(\boldsymbol{\theta}^{(1)}), .., g(\boldsymbol{\theta}^{(B)})$.

4.6 Scoring algorithm

The scoring algorithm is a Newton-type method for solving maximum likelihood equations numerically by iteratively estimating a root of the first derivative of the log-likelihood. Assume that the model parameters are given by the vector $\boldsymbol{\theta} \in \mathbb{R}^K$. The first-order derivative of the log-likelihood is called the *score function* and is denoted by the vector $\mathbf{S}(\boldsymbol{\theta}) \in \mathbb{R}^K$. The matrix with the negative of the second-order partial derivatives is called the *observed information matrix* and is denoted by the $K \times K$ matrix $\mathbf{I}(\boldsymbol{\theta})$.

For a given $\boldsymbol{\theta}_0$, the Newton-Raphson method works with the first-order Taylor expansion about $\boldsymbol{\theta}_0$; that is, with

$$\mathbf{S}(\boldsymbol{\theta}) \approx \mathbf{S}(\boldsymbol{\theta}_0) - \mathbf{I}(\boldsymbol{\theta}_0)(\boldsymbol{\theta} - \boldsymbol{\theta}_0).$$

For the maximum likelihood estimate denoted $\widehat{\boldsymbol{\theta}}$, we have $\mathbf{S}(\widehat{\boldsymbol{\theta}}) = 0$ and it follows that

$$\widehat{\boldsymbol{\theta}} \approx \boldsymbol{\theta}_0 + \mathbf{I}(\boldsymbol{\theta}_0)^{-1}\mathbf{S}(\boldsymbol{\theta}_0).$$

This is the motivation to estimate the maximum of the likelihood iteratively by

$$\boldsymbol{\theta}_{m+1} = \boldsymbol{\theta}_m + \mathbf{I}(\boldsymbol{\theta}_m)^{-1}\mathbf{S}(\boldsymbol{\theta}_m), \tag{4.9}$$

for $m = 0, 1, 2, ...$, and with $\boldsymbol{\theta}_0$ a vector with suitable starting values.

A modification often used in applications is replacing the observed information matrix by its expected value. The resulting algorithm is called the *Fisher scoring algorithm*.

Kalbfleisch and Lawless (1985) use this modification to define a scoring algorithm for continuous-time multi-state models. The likelihood for multi-state survival model (4.7) is different from the likelihood in Kalbfleisch and Lawless (1985). The summation in (4.7) is over individual data instead of aggregated data, and (4.7) includes contributions for right-censored states and known death times. The summation for individual data is also presented in Gentleman et al. (1994). These differences induce adaptations of the scoring algorithm in Kalbfleisch and Lawless (1985) but do not change the overall approach.

The main difference with the work by Kalbfleisch and Lawless (1985) lies in the scope of using the algorithm for models where the piecewise-constant approximation of parametric shapes is adopted. When the algorithm is implemented such that parametric shapes can be transition-specific, it becomes a powerful tool for flexible statistical modelling

To specify the scoring algorithm, we present the derivative of the log-likelihood first. Given piecewise-constant intensities, the likelihood contribution for an observed time interval $(t_1, t_2]$ is defined using a constant generator matrix $\mathbf{Q} = \mathbf{Q}(t_1|\boldsymbol{\theta})$, where $\boldsymbol{\theta}$ is the vector with the model parameters. The k entry of $\boldsymbol{\theta}$ will be denoted by θ_k. For the eigenvalues of $\mathbf{Q}$ given by $\mathbf{b} = (b_1, ..., b_D)$, define $\mathbf{B} = \text{diag}(\mathbf{b})$. Given matrix $\mathbf{A}$ with the eigenvectors as columns, the eigenvalue decomposition is $\mathbf{Q} = \mathbf{ABA}^{-1}$. Transition probability matrix for elapsed time $t = t_2 - t_1$ is now given by

$$\mathbf{P}(t) = \mathbf{A}\ \text{diag}\left(e^{b_1 t}, ..., e^{b_D t}\right)\ \mathbf{A}^{-1}.$$

As described in Kalbfleisch and Lawless (1985), the derivative of $\mathbf{P}(t)$ can be obtained as

$$\frac{\partial}{\partial \theta_k}\mathbf{P}(t) = \mathbf{A}\mathbf{V}_k\mathbf{A}^{-1},$$

where $\mathbf{V}_k$ is the $D \times D$ matrix with (l,m) entry

$$\begin{cases} g_{lm}^{(k)} \left[\exp(b_l t) - \exp(b_m t)\right]/(b_l - b_m) & l \neq m \\ g_{ll}^{(k)} t \exp(b_l t) & l = m, \end{cases}$$

where $g_{lm}^{(k)}$ is the (l,m) entry in $\mathbf{G}^{(k)} = \mathbf{A}\partial\mathbf{Q}/\partial\theta_k\mathbf{A}^{-1}$. This result can also be found in Jennrich and Bright (1976).

Matrix $\partial\mathbf{Q}/\partial\theta_k$ is in most cases easy to derive, which makes calculation of $\partial\mathbf{P}(t)/\partial\theta_k$ relatively easy. In the time-homogeneous case, $\mathbf{Q} = \mathbf{Q}(t_1|\boldsymbol{\theta})$ is not dependent on t_1. Changing to a piecewise-constant time-dependent model with transition-specific baseline distributions induces changes of $\partial\mathbf{Q}/\partial\theta_k$ due to dependence on t_1, but does not affect the basic steps of the scoring.

Next, we present the scoring algorithm. To do this succinctly, the constituting parts in the individual contribution to the likelihood (4.7) are denoted by

$$L_{ij} = \begin{cases} P(Y_j = y_j | Y_{j-1} = y_{j-1}, \boldsymbol{\theta}, \mathbf{x}) & \text{for} \quad j = 2, \ldots, J_i - 1 \\ C(y_{J_i}, y_{J_i - 1}, \boldsymbol{\theta}, \mathbf{x}) & \text{for} \quad j = J_i. \end{cases}$$

The k entry of the score $\mathbf{S}(\boldsymbol{\theta})$ is thus given by

$$\mathbf{S}_k(\boldsymbol{\theta}) = \sum_{i=1}^{N}\sum_{j=2}^{J_i} \frac{\partial}{\partial\theta_k} \log L_{ij}.$$

The expected observed information matrix is called the *Fisher information* and it is given by

$$\mathcal{I}(\boldsymbol{\theta}) \quad = \quad \mathbb{E}\left[\mathbf{S}(\boldsymbol{\theta})\mathbf{S}(\boldsymbol{\theta})^\top\right].$$

The asymptotic covariance matrix of the maximum likelihood estimate $\widehat{\boldsymbol{\theta}}$ is equal to $\mathcal{I}(\boldsymbol{\theta})^{-1}$. To estimate the Fisher information, define the (k,l) entry of $\mathbf{M}(\boldsymbol{\theta})$ by

$$\sum_{i=1}^{N}\sum_{j=2}^{J_i} \frac{\partial}{\partial\theta_k} \log L_{ij} \frac{\partial}{\partial\theta_l} \log L_{ij},$$

for $k,l \in \{1,...,K\}$. With $\mathbf{S}(\boldsymbol{\theta})$ and $\mathbf{M}(\boldsymbol{\theta})$ defined for any value of $\boldsymbol{\theta}$, the maximum likelihood estimate can be iteratively estimated using the algorithm

$$\boldsymbol{\theta}_{m+1} = \boldsymbol{\theta}_m + \mathbf{M}(\boldsymbol{\theta}_m)^{-1}\mathbf{S}(\boldsymbol{\theta}_m),$$

which is an estimation of (4.9).

After convergence, the covariance matrix of the maximum likelihood estimate $\mathbf{M}(\boldsymbol{\theta})$ is estimated by $\mathbf{M}(\widehat{\boldsymbol{\theta}})^{-1}$.

4.7 Model comparison

Throughout this book, the Akaike's information criterion (AIC, Akaike, 1974) is used to compare and select models. The criterion is given by

$$AIC = -2\log(L_{\max}) + 2k, \tag{4.10}$$

where $L_{\max}$ is the likelihood value at the maximum likelihood estimate, and k is the number of independent parameters in the model.

We will not discuss the theory of the AIC in detail. For more information on the AIC and other criteria see Glaeskens and Hjort (2008). It is also this publication upon which the following summary is based.

Including more parameters in a model will always decrease the value of $-2\log(L_{\max})$. The penalty $2k$ in (4.10) is to avoid models that are too complex in the sense of containing too many parameters. We will call the model with the fewest parameters the most parsimonious model.

An alternative method for comparing models is using the Bayesian information criterion (BIC, Schwartz, 1978) which is given by

$$BIC = -2\log(L_{\max}) + \log(n)k, \tag{4.11}$$

where n is the sample size of the data. The BIC has a stronger penalty than the AIC for $n \geq 8$. But there is a problem with respect to the choice of n. For longitudinal data, one can choose n to be the number of individuals, or n equal to the total number of observations. The former is the most common choice; see, for example, Muthén and Asparouhov (2009). The definition of the BIC is for n equal to the total number of independent observations, hence both of the above choices are not optimal. See Carlin and Louis (2009, Section 4.6.1) for a discussion of this issue and further references.

If we distinguish the density defined by the model, say $f(y|\theta)$, and the true but unknown density of the data, say $g(y)$, then the distance between the parametric f and the true g can be defined by the Kullback-Leibler (KL) distance given by $KL(f,g) = \int g(y)\log\left(g(y)/f(y|\theta)\right)dy$.

If there is a set of models that includes exactly one model with the minimum KL distance, then both AIC and BIC will select this model. In other words, both criteria are weakly consistent; that is, with probability tending to one as the sample size goes to infinity, the AIC and the BIC are both able to select the model with the minimum KL distance to the true model. However, if there is more than one model with the minimum KL distance, AIC will not necessarily select the most parsimonious model whereas the BIC will (Glaeskens and Hjort, 2008, Section 4.1).

An alternative to selecting a model using consistency is to select a model on the basis of efficiency, that is, with respect to minimising a loss function. The loss function can be defined, for instance, by the expected squared predicted error. For this loss function, Glaeskens and Hjort (2008, Section 4.6 and 4.7) show that in the context of autoregressive models and normal linear regression, the AIC is asymptotically efficient and the BIC is not.

If the true model is in the set of candidate models, the BIC will select this model whereas the AIC might not. However, such a situation is not to be expected in the practice of longitudinal data analysis, where models tend to be too simple to be considered the true model for the process of interest. Because prediction is often an important aim for fitting multi-state models, we will use the AIC as the default criterion in this book.

The above is about comparing candidate models. It is not about how well the chosen model fits the data in an absolute sense. If all the candidate models fit poorly, the criteria will help to choose the best among the bad models, which is—however—still a bad model. Goodness of fit, or more general, model validation, is discussed in Section 4.9.

The AIC and the BIC are criteria which can be used when the estimation is by maximum likelihood. In Chapter 5, Bayesian inference will be discussed for multi-state survival models including the deviance information criterion (DIC) for model comparison.

4.8 Example

4.8.1 English Longitudinal Study of Ageing (ELSA)

In many investigations in biostatistics, survival is not of primary interest but it may be that statistical modelling has to take into account dropout due to death because it is associated with the process of interest. For example, dropout due to death cannot be ignored when older people are followed up with respect to a process that is associated with ageing. In such a situation there is a need for a joint model for the process of interest *and* survival.

To illustrate we discuss and analyse longitudinal data from the English Longitudinal Study of Ageing (ELSA, `www.ifs.org.uk/ELSA`). The ELSA baseline (1998-2001) is a representative sample of the English population aged 50 and older. ELSA contains information on health, economic position, and quality of life. Longitudinal data on cognitive function are available in the waves 1-5 (2002-2011). Data from ELSA can be obtained via the Economic and Social Data Service (`www.esds.ac.uk`). There are 11,932 individuals in the ELSA baseline.

In addition to the so-called *core sample members* in ELSA, cohabiting spouses or partners of core sample members are also included in ELSA. This inclusion is irrespective of the age of the spouse or partner and because of this there are individuals who were younger than 50 at baseline wave 1.

For the analyses in this book, we use a sample of size $N = 1000$. This subsample is used to illustrate the statistical methods for multi-state models—results should not be used to inform clinical practice.

Restrictions were imposed for the subsample. The first is being 50 years or older at the ELSA baseline. Because ELSA data are publicly available, measures have been taken by the data provider to prevent identification of the individuals. One of those measures is the censoring of ages above 90 years. The second restriction for the subsample is being younger than 90 years at baseline without a censored age during follow-up. The third restriction is having at least two follow-up times. The latter requirement is also met if there is a baseline interview followed up by a time of death or time of right censoring.

The $N = 1000$ individuals in the subsample were randomly selected from the individuals who met the requirements. The resulting sample has 544 women and 456 men. Frequencies for number of observations per individual (including death) are 194, 140, 158, 497, and 11 for number of observations 2, 3, 4, 5, and 6, respectively.

Highest educational qualification is dichotomised for the current analysis: the higher level implies an education up to at least Level 2 of the National Vocational Qualification, or a General Certificate of Education at Ordinary Level, or an equivalent. With this the higher level implies ten or more years of formal education. There are 558 individuals with the lower education level and 442 with the higher.

During follow-up, 205 of the 1000 individuals die. A dropout rate of 20% is too much to ignore in the data analysis, especially if the process of interest is associated with ageing. For ageing processes that can be described by trajectories of transitions in a finite set of living states, death can be taken info account by adding dead as an absorbing state. This defines a multi-state survival model. The next section will give an example.

Table 4.1 *State table for the ELSA data: number of times each pair of states was observed at successive observation times. The four living states are defined by number of words remembered.*

	To				
From	10-7 words	6-5 words	4-2 words	1-0 words	Dead
10-7 words	164	150	49	12	8
6-5 words	156	440	303	48	40
4-2 words	52	336	616	151	85
1-0 words	11	35	114	149	72

4.8.2 *A five-state model for remembering words*

Of interest is change of cognitive function over time. There are a number of questions in ELSA that concern cognitive function. Here we focus on the number of words remembered in a delayed recall from a list of ten: "A little while ago, you were read a list of words and you repeated the ones you could remember. Please tell me any of the words that you can remember now." The score on this test is equal to the number of words remembered; that is, score $\in \{0, 1, 2, ..., 10\}$. The top panel of Figure 4.1 provides information on the number of words remembered at baseline. Most people remember 4 or 5 words, and the data show that remembering 9 or 10 words is exceptional. The bottom panel of Figure 4.1 depicts the change of number of words remembered over time for a random subset of 40 individuals. The 40 trajectories illustrate that the delayed recall is a noisy process. Nevertheless, already in the depicted trajectories there is some evidence of a decline in cognitive function when people get older. The statistical modelling in this section explores the effect of age and gender on cognitive change over time when controlling for education.

Four living states are defined by the number of words an individual can remember: state 1, 2, 3, and 4, for the number of words $\{7, 8, 9, 10\}$, $\{6, 5\}$, $\{4, 3, 2\}$, and $\{1, 0\}$, respectively; see also Figure 4.2.

The interval-censored multi-state process is summarised by the frequencies in Table 4.1. Note that the sum of the transitions into the dead state is equal to the number of deaths in the sample; that is, 205. The diagonal of the 4×4 subtable for the living states dominates. This shows that if there is change over time, then this change is slow relative to the follow-up times in the ELSA study. Table 4.1 also shows that the process is mainly progressive in the sense that the main trend over time is toward the higher states.

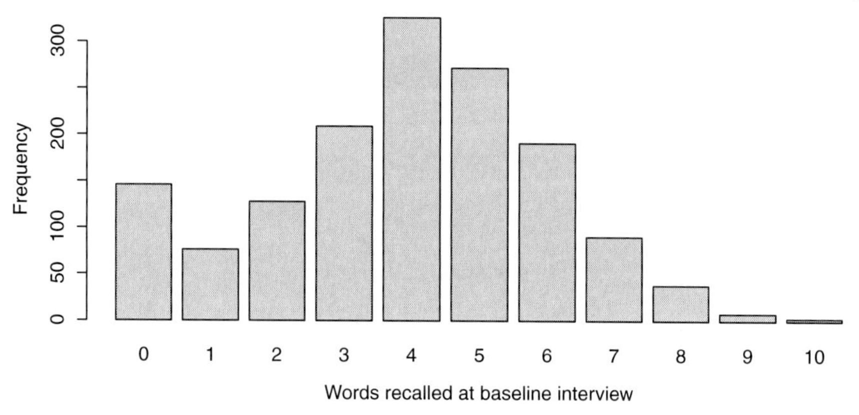

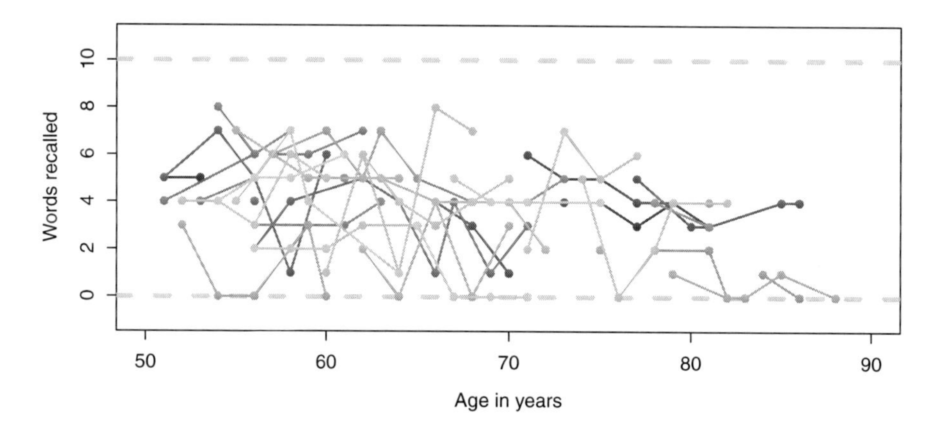

Figure 4.1 *Number of words remembered at the ELSA baseline (top panel), and follow-up trajectories for a random subset of 40 individuals (bottom panel).*

In what follows, model estimation is undertaken by using the scoring algorithm (with tolerance $\varepsilon = 1 \times 10^{-6}$). As always with numerical optimisation, it is good to explore multiple sets of starting values. In what follows, only one set of starting values is reported. Model selection is bottom-up starting with the time-homogeneous exponential hazard model given by

$$q_{rs}(t) = \exp\left(\beta_{rs.0}\right),$$

for $(r,s) \in \{(1,2),(1,5),(2,1),(2,3),(2,5),(3,2),(3,4),(3,5),(4,3),(4,5)\}$. This intercept-only model with 10 parameters has AIC = 8109.513. Convergence of the scoring algorithm was reached after 14 iterations, using starting values $\beta_{rs.0} = -3$ for all the parameters.

For the process at hand, age is the most suitable time scale. Age in the ELSA data is transformed by subtracting 49 years. This results in 1 being

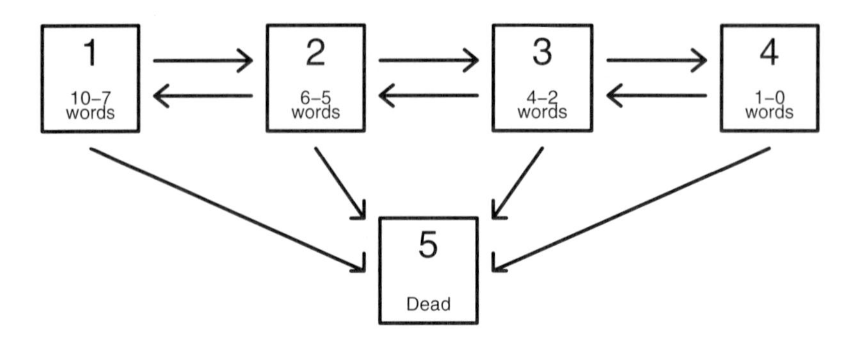

Figure 4.2 *Five-state model for survival and number of words remembered in ELSA.*

the minimal age in the sample. Even though the sample size is not small, Table 4.1 shows that mortality information is limited because only 20% of the individuals end up in the dead state during follow-up. For example, amongst those who die during the follow-up, only 8 individuals have state 1 as the last observed living state.

The following extends the model by adding parameters for progressive transitions only, and by imposing parameter equality constraints for the transitions into the dead state.

Using the piecewise-constant approximation for parametric hazard models, a Gompertz model is fitted with restrictions on the parameters for the effect of age. The grid for the piecewise-constant approximation is defined by individually observed follow-up times. The model is given by

$$q_{rs}(t) = \exp\left(\beta_{rs.0} + \xi_{rs}t\right), \tag{4.12}$$

where $\xi_{21} = \xi_{32} = \xi_{43} = 0$ and $\xi_{15} = \xi_{25} = \xi_{35} = \xi_{45}$. This model has 14 parameters, and needs 17 scoring iterations when using starting values $\beta_{rs.0} = -3$ and $\xi_{rs} = 0$ for all the relevant r,s-combinations. The model has AIC = 7805.0.

Next, covariate information is added for the transitions of interest, that is, for those transitions that represent a decline in cognitive function. For this, model (4.12) is extended by

$$q_{rs}(t) = \exp\left(\beta_{rs.0} + \xi_{rs}t + \beta_{rs.1}\ sex + \beta_{rs.2}\ education\right), \tag{4.13}$$

where *sex* is 0/1 for women/men, and *education* is 0/1 for the lower/higher level of education. For the transitions into the dead state, the constraints on

Table 4.2 *Comparison between models for the ELSA data with* $N = 1000$, *using the maximum value of the log-likelihood function and the AIC.*

Model	Baseline hazards	#Parameters	$-2\log(L_{\max})$	AIC
Intercept-only	Exponential	10	8089.5	8109.5
t	Gompertz	14	7777.0	7805.0
t, sex, $education$	**Model I**: Gompertz	21	7661.4	7703.4
t, sex, $education$	**Model II**: Gompertz and Weibull for death	21	7670.2	7712.2
t, sex, $education$	Weibull	21	7698.0	7740.0
t, sex, $education$	Weibull and Gompertz for death	21	7687.9	7729.9

the coefficients for *sex* are $\beta_{15.1} = \beta_{25.1} = \beta_{35.1} = \beta_{45.1}$, and for *education* we set $\beta_{r5.2} = 0$ for $r = 1, 2, 3, 4$. This model has 21 parameters, needs 17 iterations, and has AIC = 7703.4.

It is worthwhile to investigate alternative time-dependent models. First, in model (4.13), the Gompertz baseline models for the transitions into the dead state are replaced by Weibull models. Starting values for the transitions into the dead state are $\beta_{r5.0} = -10$, $\tau_{15} = \exp(0.5)$, and for the remaining parameters the values are as given above. This yields AIC = 7712.2 after 21 iterations.

Next, all baseline hazard definitions in model (4.13) are replaced by Weibull models, which results in AIC = 7740.0 after 31 iterations. Alternatively, model (4.13) is defined with Gompertz baseline models for the transitions into the dead state and Weibull models for progression through the living states. This yields AIC = 7729.9 after 27 iterations.

Table 4.2 summarises the comparison of the models so far. It is clear that choosing Weibull baseline hazards for the progression through the living states is not supported by the AICs. The model with the lowest AIC is the one with Gompertz baselines for all the progressive transitions (Model I). The model with Gompertz for the progression through the living states and Weibull for death (Model II) has an AIC which is very close to the lowest AIC.

For the model and data at hand, the scoring algorithm is robust in the sense that if the starting values yield a value of the log-likelihood, then the additional iterations of the algorithm lead to the same solution. Number of required iterations may vary quite a bit, however, across various sets of starting values. For Model I, it varied from 17 up to 28 iterations. Numerical problems at the first iteration occurred with starting values which include large positive values for (some of the) ξ-parameters. Because of the use of the exponential function in the model, these large values can lead to numerical problems. The best strategy for the starting values is to use negative values for the intercepts and to set the values for the other parameters to zero.

To recap, Model I is given by

$$\begin{aligned} q_{rs}(t) &= \exp\left(\beta_{rs.0} + \xi_{rs}t + \beta_{rs.1}\ sex + \beta_{rs.2}\ education\right) \\ &\qquad \text{for } (r,s) \in \{(1,2),(2,3),(3,4)\} \\ q_{rs}(t) &= \exp\left(\beta_{rs.0}\right) \\ &\qquad \text{for } (r,s) \in \{(2,1),(3,2),(4,3)\} \\ q_{rs}(t) &= \exp\left(\beta_{rs.0} + \xi_D t + \beta_{D.1}\ sex\right) \\ &\qquad \text{for } (r,s) \in \{(1,5),(2,5),(3,5),(4,5)\}. \end{aligned} \tag{4.14}$$

For Model II, the model for the transition into the dead state is changed to

$$\begin{aligned} q_{rs}(t) &= \tau_D t^{(\tau_D - 1)} \exp\left(\beta_{rs.0} + \beta_{D.1}\ sex\right) \\ &\qquad \text{for } (r,s) \in \{(1,5),(2,5),(3,5),(4,5)\}, \end{aligned} \tag{4.15}$$

Parameter estimates for Model I are presented in Table 4.3. Most of the point estimates are according to expectation. For example, the effect of getting older is associated with decline of cognitive function $\widehat{\xi}_{12}, \widehat{\xi}_{23}, \widehat{\xi}_{34} > 0$. For transitions $1 \to 2$, $2 \to 3$, and $3 \to 4$ more education is associated with a lower risk of moving.

Figure 4.3 shows the time dependency of the transition intensities for men and women conditional on a higher level of education ($education = 1$). The graphs clearly show the difference between men and women, with men having a higher risk of cognitive decline and death. There is an increase of risk of progression over the years to a decline of cognitive function, but the graphs also show that this is partly counterbalanced by high transition intensities for $3 \to 2$ and $4 \to 3$. Although graphs of transition intensities help to understand the estimated model, interpretation is more straightforward when we consider transition probabilities.

Firstly, consider a short time interval for which we assume that the intensities are constant. For men aged 60 with higher level of education, the 2-year transition probabilities are estimated at

$$\widehat{\mathbf{P}}\left(\begin{matrix} t_1 = 11, \\ t_2 = 13 \end{matrix} \middle| \begin{matrix} sex = 1, \\ education = 1 \end{matrix}\right) = \begin{pmatrix} 0.335 & 0.493 & 0.147 & 0.007 & 0.017 \\ 0.173 & 0.539 & 0.249 & 0.017 & 0.021 \\ 0.077 & 0.373 & 0.460 & 0.059 & 0.031 \\ 0.025 & 0.165 & 0.370 & 0.394 & 0.046 \\ 0 & 0 & 0 & 0 & 1 \end{pmatrix},$$

Table 4.3 *Parameter estimates for the five-state Model I for the ELSA data on cognitive function. Estimated standard errors in parentheses. Time scale t is age in years minus 49.*

Intercept			t			*sex*		
$\beta_{12.0}$	−1.133	(0.189)	ξ_{12}	0.052	(0.009)	$\beta_{12.1}$	0.543	(0.137)
$\beta_{15.0}$	−6.338	(0.862)	ξ_{23}	0.084	(0.011)	$\beta_{23.1}$	0.477	(0.151)
$\beta_{21.0}$	−1.357	(0.097)	ξ_{34}	0.047	(0.006)	$\beta_{34.1}$	0.192	(0.100)
$\beta_{23.0}$	−1.060	(0.152)	ξ_D	0.063	(0.008)	$\beta_{D.1}$	0.139	(0.145)
$\beta_{25.0}$	−6.064	(0.443)						
$\beta_{32.0}$	−0.774	(0.078)				*education*		
$\beta_{34.0}$	−2.984	(0.223)				$\beta_{12.2}$	−0.291	(0.144)
$\beta_{35.0}$	−5.429	(0.359)				$\beta_{23.2}$	−0.828	(0.102)
$\beta_{43.0}$	−0.749	(0.101)				$\beta_{34.2}$	−0.443	(0.160)
$\beta_{45.0}$	−5.001	(0.437)						

where t denotes age transformed by subtracting 49 years. The diagonal entries in this matrix dominate in the sense that within each row r, the probability p_{rr} is the largest. Nevertheless, there are some large off-diagonal entries as well. For example, a man aged 60 in state 3 has a 37% chance of being in state 2 after 2 years. This high conditional chance is an illustration of the noisiness of the process under investigation: it is quite likely that a 60 year old moves between states 2 and 3 within the next 2 years.

Next we illustrate the estimation of standard errors and 95% confidence intervals for transition probabilities. Rows in transition probability matrices sum up to 1, and—for the multi-state survival model—the last row is trivial and without uncertainty. Because of this the delta method is utilised for the estimation of the submatrix defined by the entries p_{rs}, for $r = 1,...,D-1$, $s = 1,...,D$. Applying this to the current case with $D = 5$, the estimated standard errors for the first four rows are given by

$$\begin{pmatrix} 0.039 & 0.030 & 0.015 & 0.001 & 0.007 \\ 0.013 & 0.020 & 0.019 & 0.003 & 0.005 \\ 0.007 & 0.017 & 0.020 & 0.009 & 0.006 \\ 0.003 & 0.014 & 0.023 & 0.033 & 0.012 \end{pmatrix}.$$

Using simulation with $B = 1000$, we obtain

$$\begin{pmatrix} 0.038 & 0.030 & 0.015 & 0.001 & 0.016 \\ 0.013 & 0.019 & 0.018 & 0.003 & 0.006 \\ 0.007 & 0.017 & 0.019 & 0.009 & 0.006 \\ 0.003 & 0.014 & 0.023 & 0.033 & 0.013 \end{pmatrix}.$$

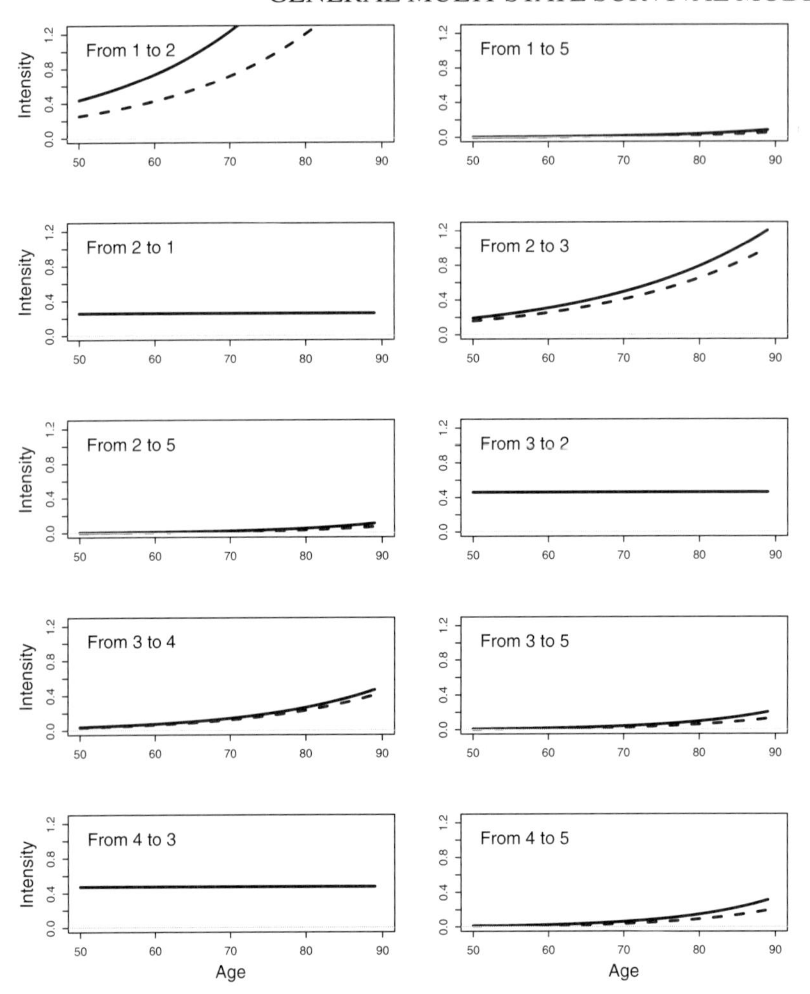

Figure 4.3 *For the 10 transitions in Model I for the ELSA data, intensities for men (solid line) and women (dashed line) given a higher level of education.*

The results for the simulation are very similar for the first four columns, but there is a clear difference for the last entry in the first row. Looking at the histograms in Figure 4.4 of simulated values of $p_{1s}(t_1,t_2)$, there is a clear difference between the symmetric distribution for $p_{1s}(t_1,t_2)$ for $s = 1,2,3,4$ and the skewed distribution for the probability of dying, that is, for $p_{15}(t_1,t_2)$. The skewness illustrates that the standard errors are of limited use for inference, and that the simulation method is the better choice as a default method. Using

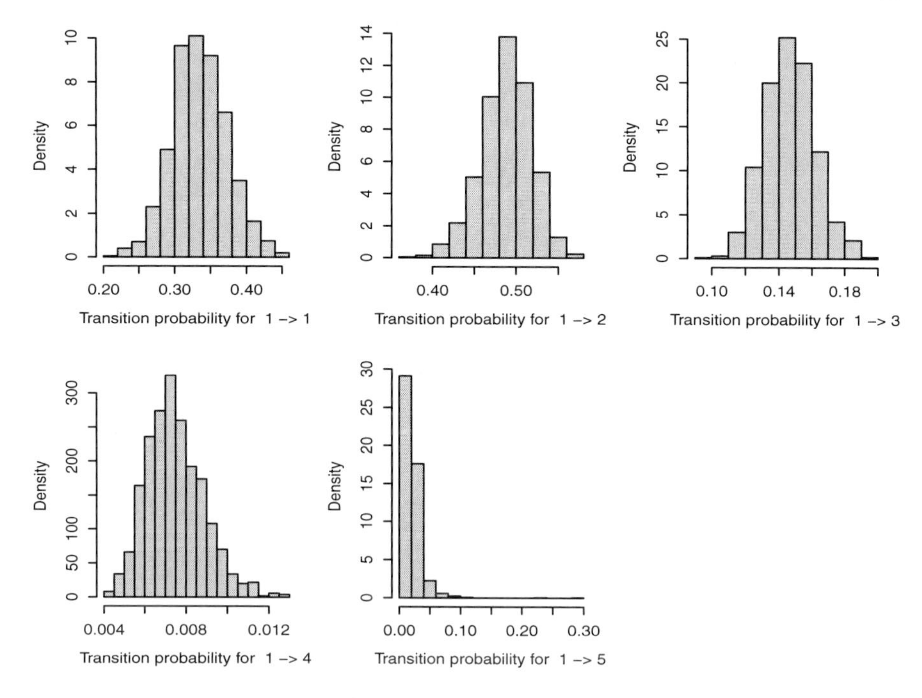

Figure 4.4 *For Model I, simulated 2-year transition probabilities for men aged 60 with higher level of education, and in state 1 at baseline. Number of replications* $B = 1000$.

simulation, the 95% confidence intervals for the first row are given by

$$(0.263, 0.410),\ (0.426, 0.543),\ (0.118, 0.179), (0.005, 0.011),\\ (0.010, 0.057).$$

Secondly, long-term prediction is illustrated for probabilities derived using a piecewise-constant approximation of the time dependency of the transition intensities. Of interest is $\mathbf{P}(t_1, t_2)$. Assume that the grid for the piecewise-constant approximation is defined by $u_{j+1} = u_j + h$ for $j = 1, ..., J$ such that $u_1 = t_1$ and $u_J = t_2$. Using the multivariate delta method, the variance of estimated $\mathbf{P}(t_1, t_2)$ is given by

$$\left(\frac{\partial \mathbf{P}(t_1, t_2)}{\partial \boldsymbol{\theta}}\right)^{\top} \widehat{\mathbf{V}}_{\theta} \left(\frac{\partial \mathbf{P}(t_1, t_2)}{\partial \boldsymbol{\theta}}\right),$$

where $\mathbf{P}(t_1, t_2) = \mathbf{P}(u_1, u_2) \times \cdots \times \mathbf{P}(u_{J-1}, u_J)$. The chain rule can be used to derive $\partial \mathbf{P}(t_1, t_2)/\partial \boldsymbol{\theta}$. For example, if $J = 3$, then

$$\frac{\partial \mathbf{P}(t_1, t_2)}{\partial \boldsymbol{\theta}} = \left(\frac{\partial \mathbf{P}(u_1, u_2)}{\partial \boldsymbol{\theta}}\right) \mathbf{P}(u_2, u_3) + \mathbf{P}(u_1, u_2) \left(\frac{\partial \mathbf{P}(u_2, u_3)}{\partial \boldsymbol{\theta}}\right).$$

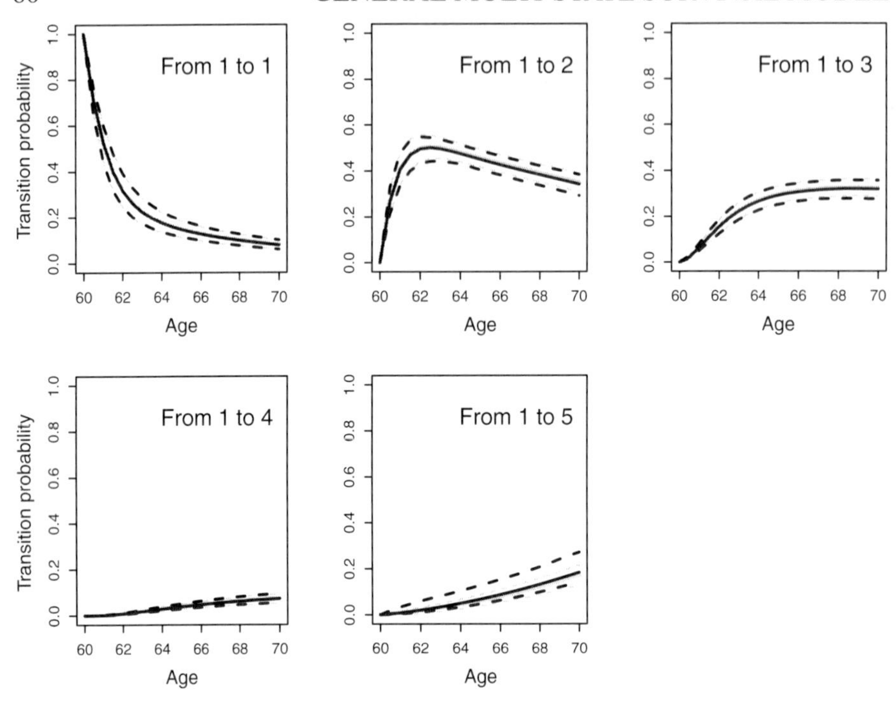

Figure 4.5 *For Model I, transition probabilities estimated up to 10 years for men aged 60 with a higher level of education, and in state 1 at baseline. Black lines for simulation method, grey for delta method. Dashed lines for 95% confidence band.*

The derivatives at the right-hand side can be computed in the same way they were computed in the scoring algorithm; see Section 4.6.

Ten-year transition probabilities are estimated for men aged 60 with 10 or more years of education. The grid is defined by $h = 1/2$ years. The estimation is shown in Figure 4.5 for both the delta method and the simulation method. For long-term prediction, the two methods produce similar results for the probability of not leaving state 1. The difference between the methods is most striking when 95% confidence bands are compared for the probability of dying.

Figure 4.5 concurs with the expectations. For example, given the progressive trend of the process, it is to be expected that probability of being in state 2 increases in the first years, but then decreases in the later years as being in states 3, 4, and 5 becomes more likely due to increased age.

Model II for the ELSA data has Weibull baseline hazards (4.15) for the transitions into the dead state instead of Gompertz hazards. The AIC of Model II is only slightly higher than the one for Model I. Estimation of transition

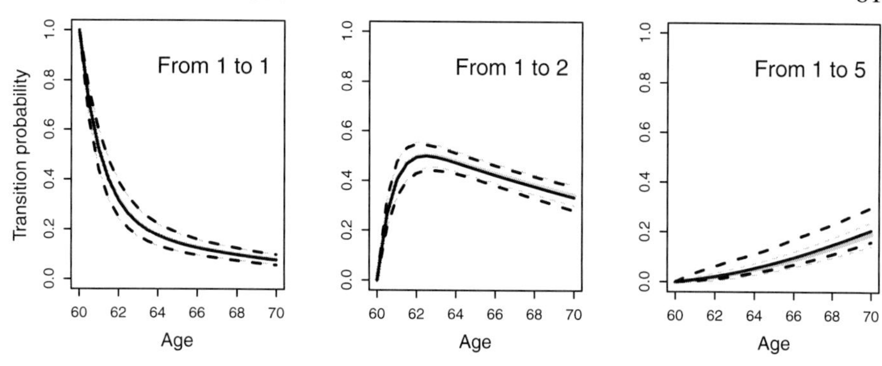

Figure 4.6 *For Model II, estimated 10-year transition probabilities for men aged 60 with a higher level of education, and in state 1 at baseline. Black lines for simulation method, grey for delta method. Dashed lines for 95% confidence band.*

probabilities is very similar for the two models. Figure 4.6 illustrates this by plotting 10-year transition probabilities predicted by Model II for men aged 60 with a higher level of education, and in state 1 at baseline. Comparing the three graphs in Figure 4.6 with the corresponding graphs in Figure 4.5 shows that the two models are quite alike with respect to the prediction. The estimate of the Weibull shape parameter is $\widehat{\tau}_D = 2.439$, with standard error 0.191. The fact that $\widehat{\tau}_D > 1$ reflects that the risk of dying increases with age. In our models, we have observed that estimation of Weibull shape parameters is often more uncertain than estimation of Gompertz shape parameters. This difference is illustrated by the 95% confidence band in the right-hand graph in Figure 4.6, which is slightly wider than the band for Model II in the last graph in Figure 4.5. For the current analysis, the difference is quite minimal.

4.9 Model validation

After a model is defined and model parameters are estimated, it is important to check whether the model fits the data. In addition, it is sometimes possible to check whether statistical inference is consistent with what is already known about the population (external validation). This section starts with discussing goodness-of-fit diagnostics for multi-state survival models. At the end of the section, external validation will be considered.

In general, goodness-of-fit assessment is difficult for multi-state survival models when censoring is present or when there is variation in observation times between and within individuals. Firstly, with interval censoring, the process is latent between observation times. A direct comparison of estimated time of transition and actual time of transition is therefore not possible. Right censoring induces a similar problem: at the time of censoring state occupancy is latent.

Secondly, when there is variation in observation times, it is not straightforward to compare the expected number of individuals in a state at a specified time with the observed number. Due to the variation, there is no guarantee that at a specified time after baseline state occupancy is observed for all individuals.

The discussion in the literature on model validation is mostly with a focus on time-homogeneous multi-state models. A good review of methods for model diagnostics is given by Titman and Sharples (2010a). The following will discuss some of these methods in the current context. Most methods are relatively easy to extend to time-dependent models.

If the model contains an absorbing state, say the dead state, for which entry times are known, then goodness of fit can be partially assessed by comparing expected survival with observed survival. Gentleman et al. (1994) suggest comparing survival estimated from the model with survival estimated by Kaplan–Meier survivor functions. This will be illustrated in Section 4.10.

Depending on the study design, it is sometimes possible to compare observed frequencies for specified time intervals with expected frequencies. For example, for balanced panel data and in the absence of known transition times, Kalbfleisch and Lawless (1985) suggest to compare observed transition frequencies with expected frequencies and formulate a Pearson chi-squared statistic. Follow-up research with attention to varying observation times and presence of covariates is presented in Aguirre-Hernandez and Farewell (2002) and Titman (2009). This is about assessing transition counts. Alternatively, and depending on the study design, one could consider prevalence counts. In the Parkinson's disease study as introduced in Section 3.7.1, the observation times of the living states do not vary much across the individuals. For the three-state model, Van den Hout and Matthews (2009) compared observed frequencies of state prevalence and expected frequencies at follow-up times. The known death times in the Parkinson's disease study are taken into account by summing the number of deaths between each pair of successive follow-up times.

When a statistic is defined to compare observed frequencies with expected frequencies, a reference distribution needs to be derived to assess the realised value of the statistic. This is where the data structure underlying multi-state survival models may create problems. Consider the comparison of observed transition frequencies with expected frequencies. For the moment assume that all individuals are observed at the same time points. Also assume that there is no right censoring. By assessing transitions, the interval censoring is not a problem. Observed successive observations in pairs of states (r, s) can be counted, and the corresponding transition probabilities can be derived from the fitted model for each individual for each time interval in the data. This allows the definition of a Pearson Chi-squared test statistic. Assume that there is a limited number of covariate values, say C. In the notation of Titman (2009), define the statistic X^2 as

$$X^2 = \sum_{j=2}^{J} \sum_{c=1}^{C} \sum_{r=1}^{D} \sum_{s=1}^{D} \frac{(o_{jcrs} - e_{jcrs})^2}{e_{jcrs}}, \tag{4.16}$$

where J is the total number of observation times, o_{jcrs} is the number of observations for time interval $(t_{j-1}, t_j]$ with state r observed at t_{j-1} and state s observed at t_j for covariate subgroup c, and e_{jcrs} is the corresponding expected number.

The definition of X^2 in (4.16) cannot be used directly when there is individual variation in observation times. Aguirre-Hernandez and Farewell (2002) present an adjusted definition of X^2 that can deal with variation in observation times and to some extent with continuous covariates. The authors present a parametric bootstrap to approximate the distribution of the test statistic. The exact distribution is intractable because under the hypothetical repetition of the process, the number of observations is not fixed due to the presence of an absorbing state; that is, the stochastic time of entry into the absorbing state determines the number of observations.

The presence of right censoring is not discussed by Aguirre-Hernandez and Farewell (2002) or Titman (2009). The implementation of the statistic in the R package msm by Jackson (2011) has functionality for taking into account right censoring, but this does not include the derivation of a p-value.

Dealing with the exact times of death in the parametric bootstrap, taking into account the variation in individual observation times, dealing with right censoring, and the discretisation of covariate values are all based upon choices that are arbitrary to some extent. Also, the extension and feasibility with respect to models with more than three states still have to be investigated.

All in all, there are a number of reasons why a Pearson Chi-squared test statistic is difficult to apply as general method. Nevertheless, comparing expected and observed frequencies (be it for transition counts or for prevalence) can be an heuristic tool for the assessment of goodness of fit. This will be illustrated in the example in the next section.

External validation of a multi-state survival model consists of comparing inference from the model with results or data from other studies. Typically this concerns summary statistics such as age-specific residual life expectancy.

These statistics are routinely provided by national bureaus of statistics. Of course, this only works if the population underlying the longitudinal study resembles the population as described by the official figures.

In the context of discrete-time multi-state survival models, Lièvre et al. (2003) analyse data from a longitudinal study on ageing and compare estimated annual probabilities of death and estimated life expectancies with information available from the U.S. National Center of Health Statistics; see also Cai et al. (2010) for a similar validation of a discrete-time model. An application of external validation of continuous-time multi-state models can be found in Van den Hout and Matthews (2008), where age-specific residual life expectancies estimated using a UK longitudinal study of ageing are compared to life expectancy estimates provided by the UK Government Actuary's Department.

4.10 Example

4.10.1 Cognitive Function and Ageing Study (CFAS)

The Medical Research Council Cognitive Function and Ageing Study (MRC CFAS, `www.cfas.ac.uk`) is a population-based longitudinal study of cognition and health conducted between 1991 and 2004 in the older population of England and Wales (Brayne et al., 2006). Individuals were recruited from six centres using an age-stratified design that oversampled those aged over 75 years at baseline. Individuals have had up to eight interviews from 1991 to 2004, and were followed since first sampled for mortality information at the Office for National Statistics. Deaths up to the end of 2005 have been included in the data that will be used here. The total sample at baseline consisted of 13,004 individuals.

CFAS is an ongoing study. Currently, CFAS II refers to new cohorts that started in 2008. The current example will only use data from CFAS conducted between 1991 and 2004. Permission to use the data for this book was kindly given by the CFAS team at the Institute of Public Health in Cambridge, UK.

One of the aims of CFAS was to investigate dementia and cognitive decline in the older population. Longitudinal data were collected by repeated interviews over the years. At interviews, background information was collected and individuals undertook several tests. One of those tests was the Mini-Mental State Examination (MMSE, Folstein et al., 1975). The MMSE is a common cognitive ability scale in population studies as it is easy to administer and measures global cognition. The MMSE integer scale is from 0 to 30. Individuals who are considered to have normal cognition score between 26 and 30, whereas individuals with severe cognitive impairment would score less than 18. Intervening groupings can also be included (Brayne and Calloway, 1990). Severe cognitive impairment would have a large impact on an individual's quality of life, since individuals in this state would have little recollection of common details, such as, for example, the current date or their address.

In this example, we use 22 as the cut-point: cognitively not impaired is defined by MMSE ≥ 22, and cognitively impaired by MMSE < 22. The value 22 is often used as the cut-point for identifying moderate and severe impairment versus mild and no impairment. According to Brayne and Calloway (1990) this cut-point is optimal for identifying mildly and more demented individuals in the age group 75-79.

The trajectory of cognition is generally downward with increasing age, however some physical illnesses can interfere with the process. For this reason, it is of interest to take into account information on stroke when investigating MMSE performance. In this section, a continuous-time multi-state survival model will be investigated for five states, where state 5 is the dead state, and living states are defined regarding cognitive function and stroke. In the absence of a history of stroke, state 1 denotes no cognitive impairment and state 2 denotes cognitive impairment. Given a history of one or more strokes, states 3 and 4 denote no cognitive impairment and cognitive impairment, respectively. Figure 4.7 shows the potential transitions between the states according to the model. Although direct transition from state 1 to state 4 is possible in practice, we opt for a parsimonious model by assuming that this transition always goes via state 3. This seems a reasonable assumption given that a continuous-time model does not impose minimal duration of stay for the living states. The last state of an individual in the data is either the dead state or a right-censored living state.

An alternative to the model in Figure 4.7 is a three-state model for cognitive function and death, with stroke as an internal time-dependent covariate. The disadvantage of such a model is that the process regarding stroke is not modelled and that prediction of the three-state process for cognition is not possible without predicting change of stroke history beforehand.

An observed improvement in cognitive function is sometimes assumed to be the result of measurement error. In a multi-state model, this would induce a misclassification of state. Misclassification will be discussed and investigated in Section 8.5.

In what follows we describe and analyse CFAS data from the Newcastle centre with 2512 individuals. At baseline there are 187 individuals with severe cognitive impairment. For this group, information on stroke history is missing or potentially unreliable. These individuals are not included in the

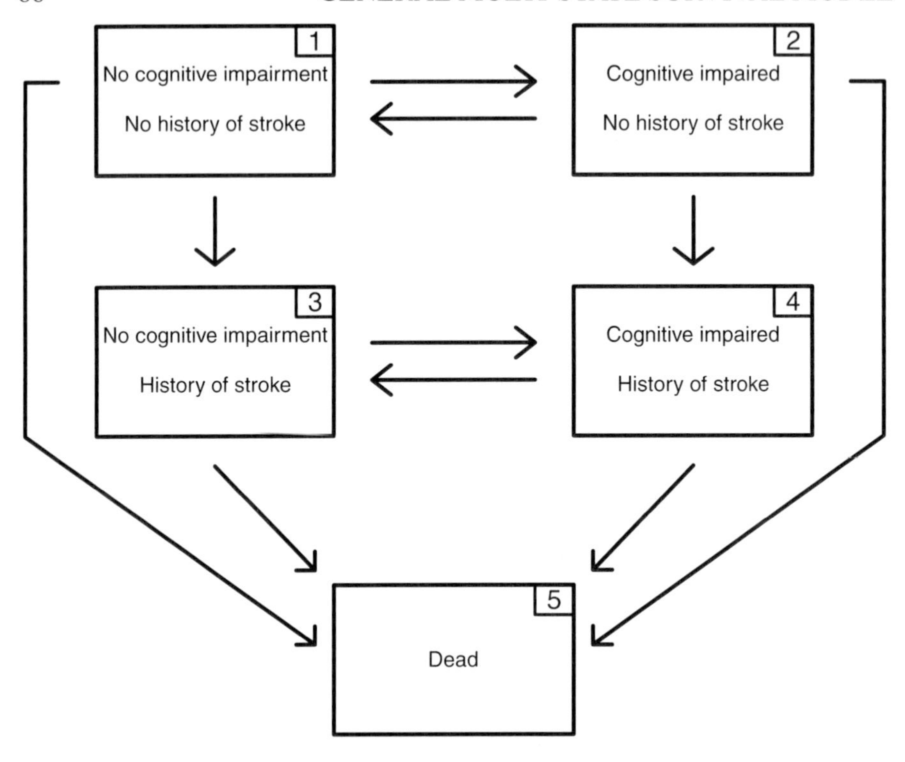

Figure 4.7 *Five-state model for cognitive function and stroke in CFAS, where no cognitive impairment is defined by MMSE* ≥ 22*, and cognitive impairment by MMSE* < 22.

analysis. Of the remaining 2325 individuals, ten have a missing state at baseline. We removed these individuals from the data as modelling missing the baseline state for such a small group is not worthwhile. Hence there are 2315 individuals (880 men; 1435 women) in the analysis. During follow-up 1550 individuals died. The total number of right-censored states is 765.

The mean length of time between interviews is 3.6 years (standard deviation 3.4). The median length of time is 2.0 years. Figure 4.8 provides information on the distribution of the observed states across age. Given that the follow-up is about 13 years and minimal baseline age in CFAS is 65, right-censored states start to appear from age 77 onward.

4.10.2 A five-state model for cognitive impairment

For the five-state process depicted in Figure 4.7, models will be specified and model validation will be illustrated. The interval-censored process is summarised by the frequencies in Table 4.4, which shows that the main trend

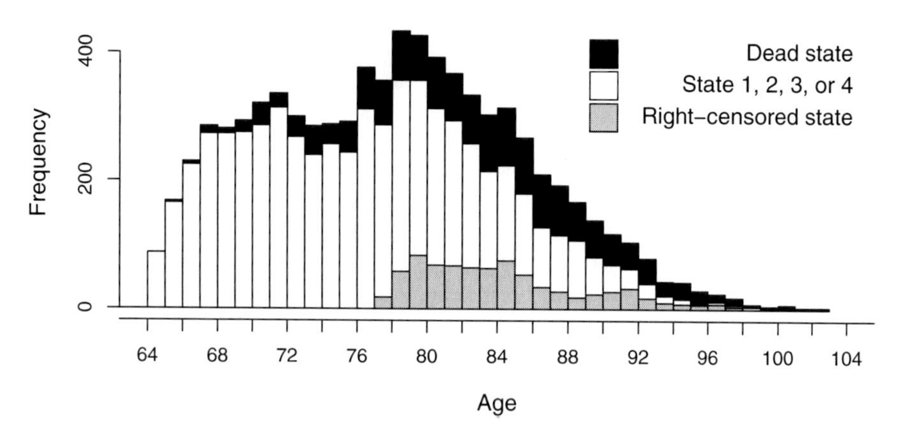

Figure 4.8 *For the five-state CFAS data, frequency counts for age at times of observation, at times of death, or at times of right censoring.*

Table 4.4 *State table for the CFAS data: number of times each pair of states was observed at successive observation times.*

	To state					
From state	1	2	3	4	Dead	Right-censored
1	2555	177	81	21	1143	670
2	69	109	1	6	188	40
3	0	0	199	30	163	49
4	0	0	16	36	56	6

over time is toward the higher states. For example, there is only one instance of an observed state 2 followed successively by an observed state 3.

Time scale is age minus 60 years. Given that the youngest individual at baseline is 64 years old, this results in 4 as the minimum age in the data analysis.

The piecewise-constant approximation is used, where the grid is defined by the data. Maximum likelihood estimation is undertaken by using the scoring algorithm. Known death times and times of right censoring are taken into account as explained in Section 4.5.

In the first model the time dependency is specified by Gompertz baseline hazards. This is Model $\mathcal{A}$ and it is given by

$$q_{rs}(t) = \exp\left(\beta_{rs.0} + \xi_{rs}t + \beta_{rs.1}sex\right), \tag{4.17}$$

with constraints

$$\xi_{15} = \xi_{25}, \quad \xi_{35} = \xi_{45}, \quad \xi_{13} = \xi_{24}, \quad \xi_{21} = \xi_{43} = 0,$$

$$\beta_{15.1} = \beta_{25.1} = \beta_{35.1} = \beta_{45.1}, \quad \beta_{13.1} = \beta_{24.1}, \quad \text{and} \quad \beta_{21.1} = \beta_{43.1} = 0.$$

The constraints are used for reason of parsimony. The constraint on $\beta_{15.1}, \beta_{25.1}, \beta_{35.1}$, and $\beta_{45.1}$, for example, implies that the effect of *sex* is the same for all four transitions into the dead state. The model has 19 parameters and AIC = 13387.9.

For a second model, the hazards for transitions to the dead state are parameterised by the Weibull baseline model, whereas the remaining transitions follow a Gompertz baseline model as above. The parameter restrictions stay the same in the sense that the restrictions on the ξ-parameters are maintained for the corresponding τ-parameters in the Weibull model. With AIC = 13389.5, this hybrid model does not perform better than Model $\mathcal{A}$.

For Model $\mathcal{A}$, Figure 4.9 provides a comparison between model-based survivor functions and Kaplan–Meier estimates of survivor functions. This comparison is baseline-state specific. Note the differences in sample sizes for the states at baseline.

There is some misfit for survival from state 4. We investigate whether parameter assumptions are too restricted and fit model (4.17) with fewer constraints on the parameters for the transition $4 \to 5$. This model is defined by

$$\xi_{15} = \xi_{25}, \quad \xi_{13} = \xi_{24}, \quad \xi_{21} = \xi_{43} = 0,$$

$$\beta_{15.1} = \beta_{25.1} = \beta_{35.1}, \quad \beta_{13.1} = \beta_{24.1}, \quad \text{and} \quad \beta_{21.1} = \beta_{43.1} = 0.$$

The model has 21 parameters and AIC = 13388.9. Given that this model does not perform better, it may be that the individuals who are in state 4 at baseline have different survival characteristics than those who are observed in state 4 during the follow-up. This seems plausible in the sense that being in state 4 and being included in the study may be associated with being less frail compared to those who are observed in state 4 during follow-up.

This assumption can be checked by adding covariate information to the model for the transition-specific intensity $q_{45}(t)$ only. We define covariate *b.state* to have value 1 if baseline state is 4, and zero otherwise. Next we fit Model $\mathcal{B}$ given by

$$q_{rs}(t) = \exp\left(\beta_{rs.0} + \xi_{12}t + \beta_{rs.1}sex + \beta_{rs.2}b.state\right), \tag{4.18}$$

where $\beta_{rs.2} = 0$ for all (r,s)-combinations other than $(r,s) = (4,5)$, and where

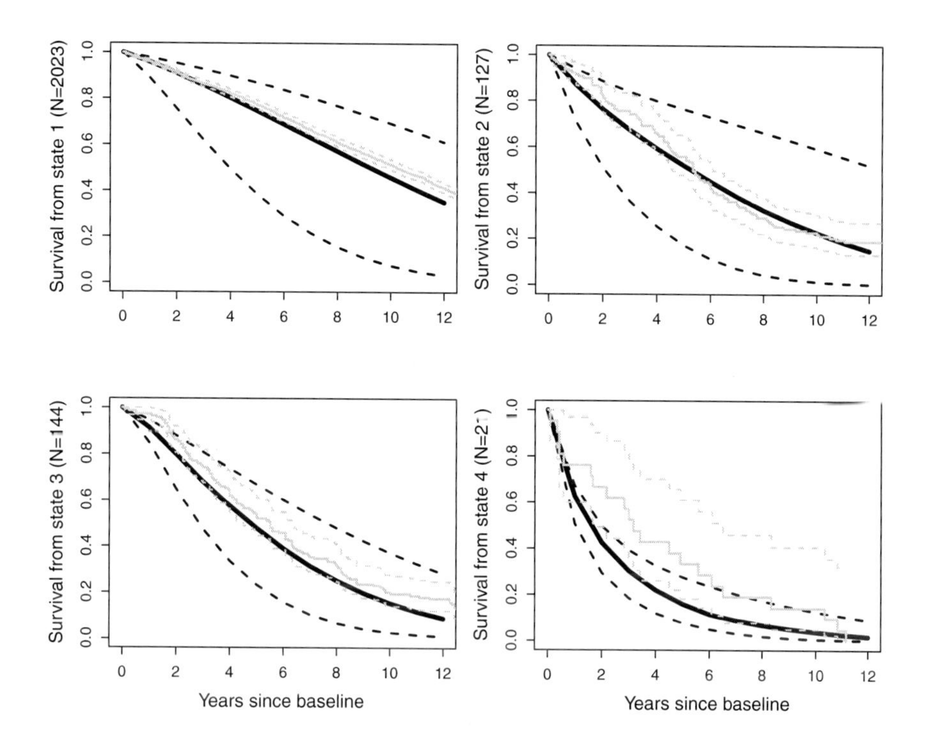

Figure 4.9 *For Model $\mathcal{A}$ for the CFAS data, comparison between model-based survival (black) and Kaplan–Meier curves (grey). State-specific graphs for the four living states at baseline (each with different sample size). Dashed lines for 95% confidence bands.*

all other constraints are the same as in Model $\mathcal{A}$. Model $\mathcal{B}$ has AIC equal to 13386.5 and Figure 4.10 shows the improved comparison between model-based survival and Kaplan–Meier curves for survival from state 4.

Using the covariate *b.state* is a way to take into account an effect that may be caused by the way the data are collected. Further investigating of this effect is outside the scope of this analysis. However, if there is indeed such a study-design effect, then the value of this covariate should be set to zero when the model is used for prediction.

The above considered model validation with respect to observed survival. If there is some indication of goodness of fit regarding observed death times, then this does not mean that the model is good at describing the transition process between the living states. For insight in goodness of fit regarding transitions between living states, a comparison of observed and expected preva-

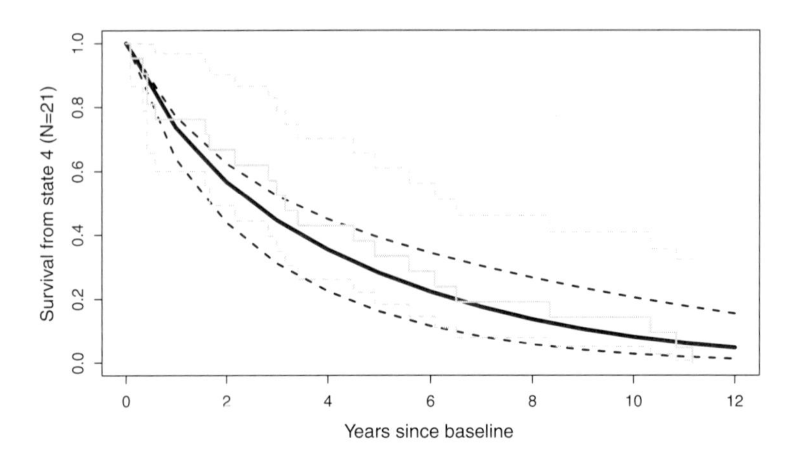

Figure 4.10 *For Model $\mathcal{B}$ for the CFAS data, comparison between model-based survival (black) and Kaplan–Meier curves (grey) for survival from state 4. Dashed lines for 95% confidence band.*

lence counts will be investigated. In this comparison, the timing of the comparison will be with respect to years since baseline.

Regarding the times at which observed and expected prevalence will be compared, we note the following. First, in the CFAS data right-censored states start occurring onward from 12 years after baseline. For this reason, the comparison of observed and expected prevalence counts will not be defined beyond the first 12 years so as to circumvent the problem of how to deal with right-censored states.

Second, the observation scheme for the CFAS is rather complex; see Figure 4.11 for the empirical distribution of follow-up times since baseline. In this figure, death times and times of right censoring are ignored. The concentration of observation times around 2, 5 and 10 years after baseline correspond with the basic study design of CFAS. As stated before, interval censoring and large variation in individual observation times hinder a direct comparison between observed and expected prevalence at pre-specified time points. To mimimise the effect of the interval censoring, the time points for the goodness-of-fit assessment are chosen to be 2, 5, and 10 years after baseline. Figure 4.11 illustrates the rationale of the choice: around these time points there are concentrations of observation times.

Given the above, the grid for the assessment of prevalence counts is $(t_1,..,t_4) = (0,2,5,10)$. Observed prevalence count for state s at time t_j, $j = 2,3,4$, is the number of individuals either in state s at t_j or with state

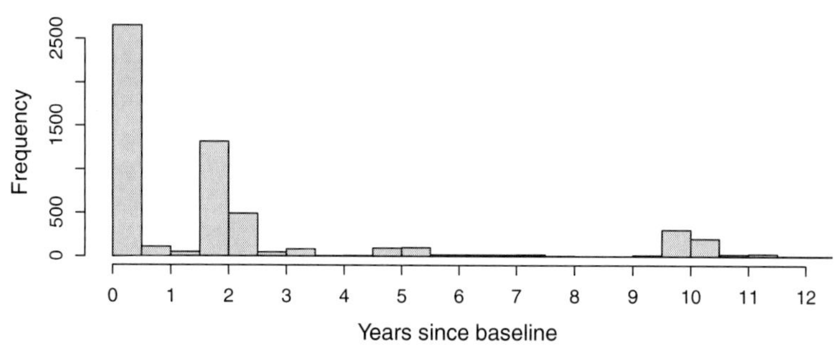

Figure 4.11 *Empirical distribution of follow-up times in the CFAS data since baseline. Only for the living states: death times and times of right censoring are ignored.*

s being the last observed state in the interval $(t_{j-1}, t_j]$. When a death is observed, then this death is counted in all the subsequent intervals. Observed prevalence counts are given by

$$\begin{pmatrix} 2023 & 127 & 144 & 21 & 0 \\ 1775 & 136 & 147 & 27 & 230 \\ 1495 & 100 & 101 & 16 & 603 \\ 961 & 61 & 65 & 8 & 1220 \end{pmatrix}, \tag{4.19}$$

where the rows correspond to the four time points, and the columns to the states; for example, no deaths at baseline, and 1220 observed deaths at 10 years after baseline. This summary of observed prevalence is compared to expected prevalence according to the fitted model.

The expected prevalence is a summary of the estimated transition probabilities for the $N = 2315$ individuals conditional on their baseline data (that is, baseline state, baseline age, and baseline covariate values). Expected prevalence counts for 0, 2, 5, and 10 years after baseline are given by

$$\begin{pmatrix} 2023 & 127 & 144 & 21 & 0 \\ 1667.5 & 164.9 & 147.2 & 62.6 & 272.9 \\ 1255.3 & 144.8 & 144.0 & 71.5 & 699.4 \\ 741.5 & 100.9 & 111.6 & 59.6 & 1301.4 \end{pmatrix}. \tag{4.20}$$

There is quite a difference between observed and expected counts. For assessing this difference, is it easier to look at the prevalence in percentages; see Figure 4.12 for a graphical representation of this comparison.

The comparison for state 5 in Figure 4.12 is basically a cruder version of the comparison in Figure 4.10 which depicts model-based survival and Kaplan–Meier curves. The graphs with the prevalence in percentages show that

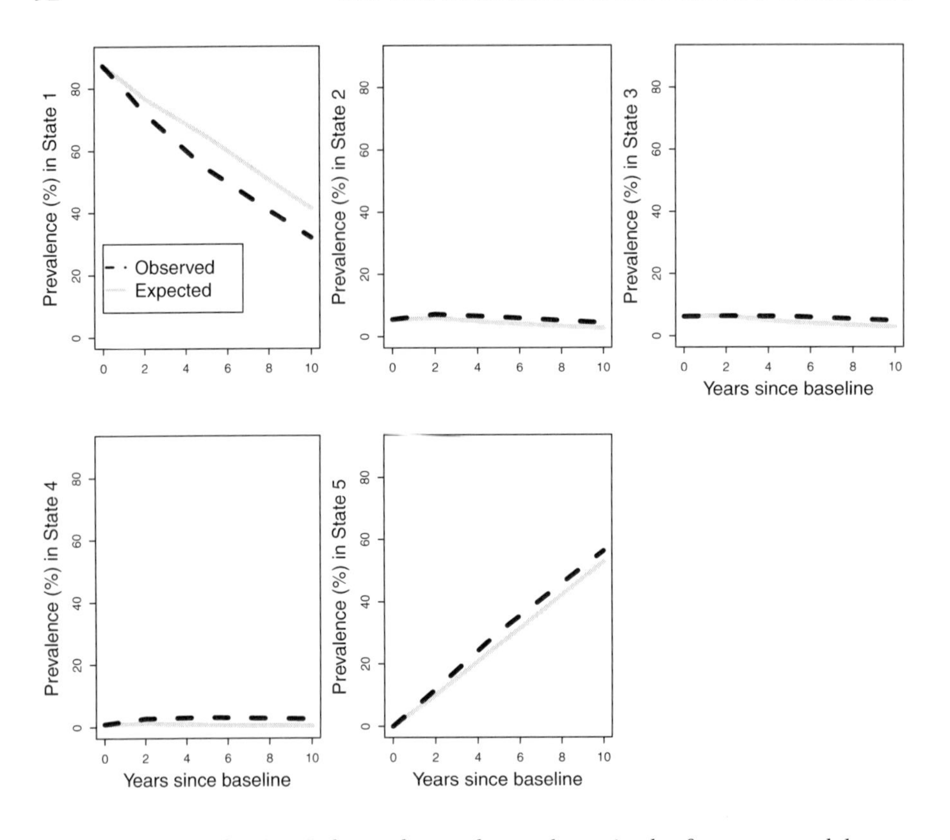

Figure 4.12 *For the CFAS data, observed prevalence in the five-state model compared with predicted prevalence. The prediction is based on Model $\mathcal{B}$ and individual data at baseline. Due to the interval censoring, the observed prevalence is an approximation (last observation of state carried forwards).*

the mapping of expected to observed seems quite good for states 2, 3, 4, and 5. There is some mismatch for state 1. The problem here is that the comparison for the living states is confounded by the interval censoring. The severity of the mismatch is therefore hard to quantify. If there is indeed a mismatch for state 1, it can be because the hazard of leaving state 1 is underestimated, but it can also be due to an overestimation of the hazard of moving from state 2 back to state 1. Further investigation of this might be undertaken by considering the comparisons for subgroups defined by baseline state. It might also be interesting to explore whether there is a correlation between the extent of the interval censoring and the state occupancy. This will not be followed-up here.

In this example, the model validation by state prevalence is used as a heuristic tool. It will pick up lack of fit for very bad models, but it is yet unclear how good the model actually is when there is no obvious lack of fit. Given the validation, there is no reason to think that the Model $\mathcal{B}$ is particularly bad. It seems to be able to capture observed survival pretty well, and is also good at predicting the state prevalence for states 2, 3, and 4 conditional on baseline data only.

Regarding the estimated model parameters (not reported here), all the coefficients for the effect of age (the ξ-parameters) are estimated at positive values illustrating the progressive nature of the process when ageing. Also according to expectations, there is a gender effect in the sense that being a man is associated with higher hazards for the transitions into the dead state. In addition, men are more likely to have a stroke. However, for the transition from state 1 to state 2, comparing men and women of the same age, women are more likely to move to the impaired state.

Chapter 5

Frailty Models

5.1 Mixed-effects models and frailty terms

Frailty models are mixed-effects models for survival data. The terminology originates from medical science, where the word frailty denotes a general susceptibility to ill health. As a random effect, a frailty term can be specified for an individual, but can also be cluster-specific in which case individuals within clusters share the same frailty. Applications of frailty models are not limited to medical science but can also be found in other disciplines where survival analysis is undertaken.

The main reason to extend the fixed-effects modelling as presented in this book so far is to take into account unobserved individual-specific or cluster-specific susceptibility for morbidity and mortality. An example of the former is a progressive multi-state survival model that uses random effects to distinguish individuals who move quickly through the states from those who do not: movers versus stayers. An example of using cluster-specific random effects is modelling survival data where patients from the same hospital share a hospital-specific frailty. If there are many hospitals in the data, then adding hospital-specific fixed effects by dummy-coded covariates will increase the number of parameters considerably which is a nuisance—especially if effects of individual hospitals are not the aim of the analysis. Another problem when fixed effects are used for clusters is that the number of parameters may increase if the sample size increases—a violation of the assumptions which allow maximum likelihood estimation.

An additional motivation for introducing frailty terms in a multi-state survival model is that frailty terms can counterbalance violations of the conditional Markov assumption. For example, in a progressive three-state illness-death model it may be the case that duration of stay in the ill-health state is correlated with the hazard of a transition to the dead state. This is a violation of the Markov assumption as defined for continuous-time processes in Section 4.2. But if a frailty is correlated with duration of stay in the ill-health state, and the hazard is modelled conditional on that frailty, then duration of stay is represented in the model even if a conditional Markov assumption is enforced.

It is possible to define a model that takes into account explicitly the correlation between duration of stay and the hazard of a transition. This is discussed in Section 8.7, which defines a semi-Markov multi-state process. Especially for interval-censored data, estimating semi-Markov models is computationally intensive. In some cases, a frailty model can be an alternative that is computational less intensive.

Specific to frailty models is the choice of the frailty distribution. Duchateau and Janssen (2008) and Aalen et al. (2008) discuss several choices for the frailty distribution in univariate survival models. Wu (2010) discusses frailty survival models within the wider framework of mixed-effects models.

Frailties can be included in univariate survival models and in multi-state survival models. This chapter only deals with the latter. In general, information about possible frailty in univariate survival data differs from information in multi-state data. Because of the repeated measurements, frailty in multi-state data may be easier to detect. Putter and Van Houwelingen (2015) provide more precise statements and details.

One specific kind of multi-state model that is problematic with respect to the identification of the frailty distribution is the competing risks model. The archetype example of a competing risks model is the three-state model, with one living state and two absorbing dead states that differ in the cause of death. Competing risks models are not featured in the current book—partly because good textbooks are already available; see, for example, Hougaard (2000) and Crowder (2012). Putter and Van Houwelingen (2015) conclude that "Identifiability of frailties in competing risks models is possible theoretically when there are covariates in the model, but is practically impossible due to the confounding of violations of the proportional hazards assumption and the presence of frailties." For multi-state processes with sequences of transitions identifiability of frailties is possible but in some cases this is only true for frailty distributions with large variances.

In what follows, a *cluster* is made up of observations that share a frailty. In the case of individual-specific frailties, the data of an individual define the cluster and there are as many clusters as there are individuals. The latter is the case in the examples in this chapter. Nevertheless, the presentation of the models is more general in the sense that clusters can also be defined by a grouping of individuals.

A note on terminology: this chapter discusses *mover/stayer models*. In these models, the frailty terms allow the distinction between individuals who move quickly through the states (the movers) and individuals who tend not to move so quickly (the stayers). Satten (1999) describes this as tracking: there is correlation between the transition times within individuals, and rapid progressors can be distinguished from slow progressors.

In the literature, a mover/stayer model can also denote a mixture model where individuals are movers when they transition through the process, or stayers when they do not move at all. More formally, this implies a stochastic process that is a mixture of two Markov processes, one of which has a transition matrix that is equal to the identity matrix; see, for example, Frydman (1984) and Cole et al. (2005). For the multi-state survival models that are featured in this book, it is not realistic to assume that there are individuals who do not move at all since the risk of mortality is never zero. Adaptations can of course be investigated; for example, a mixture model where a subset of the individuals are stayers in the sense that they do not move across the living states. This, however, will not be explored in the current chapter.

5.2 Parametric frailty distributions

In the context of a multi-state survival model, the frailty-specific hazard model for cluster c can be defined by adding a cluster-specific multiplicative term to the fixed-effects model (4.6). For transition $r \rightarrow s$, this yields the model

$$q_{rs}(t|c,\mathbf{x}) = q_{rs.0}(t)U_{rs.c}\exp\left(\boldsymbol{\beta}_{rs}^{\top}\mathbf{x}\right), \tag{5.1}$$

where $U_{rs.c}$ is a random variable. Given that (5.1) is a model for the hazard, the requirement for $U_{rs.c}$ is that it has positive support, that is, $U_{rs.c} \geq 0 \; \forall c$.

Univariate frailty distributions that will be used in this book are

$$\text{log-normal frailty:} \quad U_{rs.c} = \exp(B_{rs.c}) \text{ with } B_{rs.c} \sim N(0, \sigma_{rs}^2) \tag{5.2}$$

$$\text{gamma frailty:} \quad U_{rs.c} \sim Gamma(\phi). \tag{5.3}$$

The advantage of using the normal distribution is that it is in line with the most common distributional choice in linear mixed-effects models, and that it enables using Gauss–Hermite quadrature in the numerical optimisation of the marginal likelihood. A disadvantage of using the normal distribution is that the mean $\mathbb{E}\left(B_{rs.c}\right) = 0$ does *not* imply $\mathbb{E}\left(U_{rs.c}\right) = 1$ for the frailty distribution because of the non-linear transformation defined by the exponential.

For random variable X, the gamma distribution with shape-parameter a and scale-parameter d is defined by the density function

$$f(x) = \frac{1}{d^a\Gamma(a)}x^{a-1}\exp(-x/d) \qquad \text{for} \qquad x > 0,$$

where $a > 0$, $d > 0$, and the gamma function Γ is given by

$$\Gamma(a) = \int_0^\infty t^{a-1} \exp(-t) dt.$$

It follows that the mean and variance of X are $\mathbb{E}(X) = ad$ and $Var(X) = ad^2$. We will only use the one-parameter gamma distribution defined by choosing the combination $a = 1/\phi$ and $d = \phi$. Note that in contrast to the normal distribution, this implies $\mathbb{E}(U_{rs.c}) = 1$.

Many extensions of frailty model (5.1) can be defined. For example, the Gompertz baseline hazard can be extended using a bivariate distribution to model a random intercept and a random slope for the effect of time. Using a bivariate normal distribution, the hazard model for transition $r \to s$ can be defined as

$$\begin{aligned} q_{rs}(t|c,\mathbf{x}) &= \exp\left(\left(\beta_{rs.0} + B^{(0)}_{rs.c}\right) + \left(\xi_{rs} + B^{(1)}_{rs.c}\right) t + \boldsymbol{\beta}_{rs}^\top \mathbf{x}\right) \\ \left(B^{(0)}_{rs.c}, B^{(1)}_{rs.c}\right) &\sim N(\mathbf{0}, \boldsymbol{\Sigma}_{rs}), \end{aligned} \tag{5.4}$$

with 2×2 covariance matrix $\boldsymbol{\Sigma}_{rs}$ to be estimated from the data. Another example is to use a bivariate distribution to model correlation between two random intercepts:

$$\begin{aligned} q_{12}(t|c,\mathbf{x}) &= \exp\left(\beta_{12.0} + B_{12.c} + \xi_{12} t + \boldsymbol{\beta}_{12}^\top \mathbf{x}\right) \\ q_{23}(t|c,\mathbf{x}) &= \exp\left(\beta_{23.0} + B_{23.c} + \xi_{23} t + \boldsymbol{\beta}_{23}^\top \mathbf{x}\right) \\ (B_{12.c}, B_{23.c}) &\sim N(\mathbf{0}, \boldsymbol{\Sigma}). \end{aligned} \tag{5.5}$$

More examples will be discussed in Section 5.5.

5.3 Marginal likelihood estimation

This section discusses estimation of frailty models by marginal likelihood. Section 5.4 presents estimation using an Expectation-Maximisation (EM) algorithm, and in Chapter 6 Bayesian inference is presented using Markov Chain Monte Carlo methods.

The basic idea of marginal likelihood is to apply the law of total probability. For continuous random variables X and Z, this law is given by

$$f_X(x) = \int_{\Omega_Z} f_{X|Z}(x|z) f_Z(z) dz,$$

where Ω_Z is the sample space of z. Using the law of total probability is recommended when dealing with $f_X(x)$ head-on is hard, while dealing with $f_{X|Z}(x|z)$ is relative easy.

Writing $u_{rs.c} = \exp(b_{rs.c})$ for frailty realisation $u_{rs.c}$, consider the conditional hazard model for transition $r \to s$ given by

$$q_{rs}(t|b_{rs.c}, \mathbf{x}) = q_{rs.0}(t) \exp\left(b_{rs.c} + \boldsymbol{\beta}_{rs}^{\top}\mathbf{x}\right). \tag{5.6}$$

Conditionally on $b_{rs.c}$, the regression model has the fixed-effects format with $b_{rs.c}$ playing the role of an additional intercept. Hence with $b_{rs.c}$ given, computation of generator matrices and the associated matrices with transition probabilities does not require new methods.

Consider the general setting where $\mathbf{b}_c$ denotes the vector with the random-effects values in the linear predictor for cluster c, and $f(\mathbf{b}_c)$ the multivariate density function for the random effects. Using the same notation as in Section 4.5 and using the law of total probability, the likelihood contribution for an individual i in cluster c is

$$\begin{aligned} L_i(\boldsymbol{\theta}|c, \mathbf{y}, \mathbf{x}) &= P(Y_J = y_J, ..., Y_2 = y_2 | Y_1 = y_1, c, \boldsymbol{\theta}, \mathbf{x}) \\ &= \int_{\Omega_{\mathbf{b}_c}} P(Y_J = y_J, ..., Y_2 = y_2 | \mathbf{b}_c, Y_1 = y_1, \boldsymbol{\theta}, \mathbf{x}) f(\mathbf{b}_c) d\mathbf{b}_c \\ &= \int_{\Omega_{\mathbf{b}_c}} \left(\prod_{j=2}^{J-1} P(Y_j = y_j | \mathbf{b}_c, Y_{j-1} = y_{j-1}, \boldsymbol{\theta}, \mathbf{x}) \right) C(y_J | \mathbf{b}_c, y_{J-1}, \boldsymbol{\theta}, \mathbf{x}) f(\mathbf{b}_c) d\mathbf{b}_c, \end{aligned} \tag{5.7}$$

where $j = 1, .., J$ refers to observation times $t_1, ..., t_J$. The definition of $C(y_J| \dots)$ in (5.7) conditional on $\mathbf{b}_c$ is analogous to $C(y_J| \dots)$ in Section 4.5. Note that for individual-specific random effects, the individual defines the cluster and $\mathbf{b}_c$ can also be denoted by $\mathbf{b}_i$.

The multi-state survival models discussed so far are transition models where the likelihood function is defined conditional on the first observed state; see Section 4.5. This conditional approach, together with the Markov assumption, implies that left truncation does not have to be taken into account. However, if the model is extended by modelling the first observed state (see Sections 8.4 and 8.5), then the marginal likelihood function for the frailty model has to be adapted accordingly. For standard survival models this is discussed in, for example, Rondeau et al. (2006).

Maximisation which includes integration can be computationally demanding. For example, the frailty models (5.4) and (5.5) both contain bivariate frailty terms. This implies that a two-dimensional integral has to be approximated for the likelihood contribution of each cluster. If the clusters are defined by the individuals, then this means that each evaluation of the marginal likelihood at a given set of parameter values involves approximation of as many integrals as there are individuals in the data.

Integrals can be approximated using numerical methods. One option for a one-dimension integral is to use Simpson's rule as discussed in Chapter 3. Another option is to use Monte Carlo integration, which is conceptually simple and easy to extend to multi-dimensional integrals. For frailty models that are defined using the normal distribution, Gauss–Hermite quadrature is the best option.

We start with the Monte Carlo integration as a general method for approximating an integral, which can be written as

$$\int_{-\infty}^{\infty} g(x) f(x) dx,$$

where f is a probability density function defined on $\mathbb{R}$. Note that the integral is equal to $\mathbb{E}_X[g(X)]$ by definition. Values $x_1, x_2, \ldots, x_m$ sampled from the distribution defined by f can be used to compute the Monte Carlo estimate as

$$\frac{1}{m} \sum_{k=1}^{m} g(x_k).$$

The extension to the multi-dimensional integration is straightforward as long as sampling from the relevant multivariate probability density function is straightforward. The approximation depends of course on the choice of m, and for small m various sets of sampled values can produce quite different approximations. Larger values of m will improve the approximation, but are computationally intensive—especially in the multi-dimensional case.

Although Monte Carlo integration has the advantage of being conceptually simple, it is not the optimal method for maximising a marginal likelihood function. Besides the intensive computation, the fact that simulation is underlying the maximisation may lead to instability, which in turn may make it hard to estimate standard errors.

An alternative is to use Gauss–Hermite quadrature. The quadrature approximates the integral

$$\int_{-\infty}^{\infty} g(x) \exp(-x^2) dx.$$

The approximation is

$$\sum_{k=1}^{m} w_k g(x_k),$$

where nodes x_k, $k = 1, \cdots, m$, are the roots of the physicists' version of the Hermite polynomial $H_m(x)$, and the weights w_k are given by

$$w_k = \frac{2^{m-1} m! \sqrt{\pi}}{m^2 \left(H_{m-1}(x_k)\right)^2}.$$

These nodes and weights are readily available in software packages for statistical computing; see Appendix C.5.

For the integral

$$\int_{\infty}^{\infty} g(x) f(x) dx,$$

where f is the density of a normal distribution $N(\mu, \sigma^2)$, the approximation is

$$\frac{1}{\sqrt{\pi}} \sum_{k=1}^{m} w_k g\left(\sqrt{2}\sigma x_k + \mu\right).$$

Bivariate Gauss–Hermite quadrature can be undertaken by using two univariate normal distributions. For this approach, the bivariate normal distribution in the log-likelihood is expressed by using two univariate normal distributions. If $Z \sim N(\mu_Z, \sigma_Z^2)$ and $X \sim N(\mu_X, \sigma_X^2)$, then the density for (z, x) is equal to

$$f_{Z|X}(z|X = x) f_X(x),$$

where

$$Z|X \sim N\left(\mu_Z + \rho(\sigma_Z/\sigma_Z)(x - \mu_X), \sigma_Z^2(1 - \rho^2)\right),$$

and ρ is the correlation of X and Z; see, for example, Casella and Berger (2002, p. 177).

A general approach for approximating a multi-dimensional integral is to use any method for univariate integration iteratively for the constituent univariate integrals in the multi-dimensional integral.

When numerical integration becomes computationally infeasible due to high-dimensional integrals, using Laplace approximation has been proposed as an alternative. There is an extended literature on using Laplace approximation for random-effects models; see, for instance, Wu (2010, Section 2.6.3) and the references therein.

5.4 Monte-Carlo Expectation-Maximisation algorithm

The EM algorithm is a widely used method for maximum likelihood estimation in case data are incomplete; see Dempster et al. (1977). Incompleteness in this context should be interpreted in a wide sense. It may be that data are missing because observation failed, but it may also be the case that observation is not possible. An example of the former in a longitudinal survey is an individual missing a scheduled interview with the consequence that for the scheduled time his or her data are missing. An example of the latter are unobserved individual-specific frailties which are used to defined the frailty models in this chapter. By viewing the latent frailties as missing data, the EM algorithm can be utilised.

The EM algorithm is an iterative method in which parameter estimates are updated in each iteration. Two appealing properties of the algorithm are its numerical stability and—given that the complete-data problem is a standard one—the use of software for complete-data analysis. Given the fast scoring algorithm in Section 4.6 for complete-data analysis, exploration of the EM algorithm for frailty models is of interest.

Each iteration of the EM algorithm consists of two steps: the *E-step* and the *M-step*. The basis of the approach is to view the frailties as missing data and to define the *complete data* as the observed multi-state data *and* the frailties. The E-step then estimates the expectation of the complete-data log-likelihood conditional on the observed data and current parameter estimates, and the M-step maximises this expectation to update the parameter estimates. The two steps are iterated until convergence of the parameter estimates.

There are many versions of the EM algorithm. This section describes the Monte-Carlo EM algorithm, which is often used for non-linear mixed-effects models; see, for example, Wu (2010). The Monte-Carlo extension refers to the E-step and can be used if this step is complex in the sense that there is no closed-form solution to the computation that is needed to prepare the M-step. This will be explained in what follows.

Sutradhar and Cook (2008) also present a Monte-Carlo EM algorithm for estimating a multi-state mixed-effects model for interval-censored data. However, Sutradhar and Cook (2008) deal with the interval censoring by treating the unknown transition times as unknown variables in the EM algorithm. Their Monte-Carlo EM algorithm generates transition times and random effects. Given the definition of the marginal likelihood function (5.7) which takes into account the interval censoring by using transition probabilities, the EM algorithm in this section is with respect to the unknown random effects only.

Assume frailty terms are defined for individuals and vector $\mathbf{b}_i$ contains the frailty values for individual i. We adapt the presentation in Wu (2010, Section 2.6.2) for the current setting. Ignoring possible dependence on covariate values in the notation, the complete-data density function for individual i is given by

$$f(\mathbf{y}_i, \mathbf{b}_i|\boldsymbol{\theta}) = f(\mathbf{y}_i|\mathbf{b}_i, \boldsymbol{\theta}, \mathbf{x}) f(\mathbf{b}_i|\boldsymbol{\theta}),$$

where, as before, $\mathbf{y}_i$ is the observed data on the states, and $\boldsymbol{\theta}$ is the vector with all the model parameters. The complete-data log-likelihood contribution of individual i can thus be written as

$$l_i^{(c)}(\boldsymbol{\theta}|\mathbf{b}_i, \mathbf{y}_i) = \log\left(f(\mathbf{y}_i|\mathbf{b}_i, \boldsymbol{\theta})\right) + \log\left(f(\mathbf{b}_i|\boldsymbol{\theta})\right).$$

At the v-th iteration of the EM algorithm, the E-step computes the conditional expectation given by

$$Q(\boldsymbol{\theta}|\boldsymbol{\theta}^{(v)}) = \mathbb{E}\left(\sum_{i=1}^{N} l_i^{(c)}(\boldsymbol{\theta}|\mathbf{b}_i, \mathbf{y}_i)\middle|\mathbf{y}_i, \boldsymbol{\theta}^{(v)}\right)$$

Because the model is non-linear in the frailty terms, expectation $Q(\boldsymbol{\theta}|\boldsymbol{\theta}^{(v)})$ does not have a closed-form expression. One way to approximate the expectation is to use a Monte Carlo method. This defines a Monte Carlo EM algorithm. At the v-th iteration of the algorithm, missing frailties $\mathbf{b}_i$ are sampled from the conditional distribution $f(\mathbf{b}_i|\boldsymbol{\theta}^{(v)})$. This results in simulated vectors $\mathbf{b}_i^{(1)}, ..., \mathbf{b}_i^{(S)}$. Next $Q(\boldsymbol{\theta}|\boldsymbol{\theta}^{(v)})$ is approximated by

$$\widetilde{Q}(\boldsymbol{\theta}|\boldsymbol{\theta}^{(v)}) = \frac{1}{S} \sum_{s=1}^{S} \sum_{i=1}^{N} l_i^{(c)}(\boldsymbol{\theta}^{(v)}|\mathbf{b}_i^{(s)}, \mathbf{y}_i).$$

The M-step is the maximisation of $\widetilde{Q}(\boldsymbol{\theta}|\boldsymbol{\theta}^{(v)})$ to obtain the update $\boldsymbol{\theta}^{(v+1)}$. Because $\widetilde{Q}(\boldsymbol{\theta}|\boldsymbol{\theta}^{(v)})$ is defined conditional on frailty values, the maximisation can be undertaken using methods for fixed-effects models.

A Monte-Carlo EM algorithm is computationally very intensive. Sampling $\mathbf{b}_i$, $i = 1, ..., N$, from the conditional distribution $f(\mathbf{b}_i|\boldsymbol{\theta}^{(v)})$ can be implemented using a Metropolis algorithm (Section 6.2). This sampling has to be done in each EM-iteration and can be time-consuming for data with large N. The choice of S and the assessment of convergence of the Monte-Carlo EM algorithm are not straightforward. And—last but not least—the EM algorithm itself does not yield an estimate of parameter uncertainty. Methods to deal with these issues have been discussed in the literature; see Wu (2010) and the references therein.

An application of the Monte-Carlo EM algorithm is illustrated in the next section using the scoring algorithm for the M-step. However, for frailty models with a limited number of frailty terms, we do not recommend the EM algorithm because of its computational complexity. For the examples in this chapter, marginal likelihood (Section 5.3) and non-parametric maximum likelihood (Section 5.6) are adequate.

5.5 Example: frailty in ELSA

The English Longitudinal Study of Ageing (ELSA) subset with sample size $N = 1000$ introduced in Section 4.8.2 will be used to illustrate analysis with multi-state frailty models. For the five-state data on cognitive function and survival, fixed-effects models were fitted using age (transformed by minus 49 years) as time scale t, and with Gompertz baseline hazards for progressive transitions. The current section starts with a similar fixed-effects model defined by

$$\begin{aligned} q_{rs}(t) &= \exp\left(\beta_{rs.0} + \xi_F t + \gamma_F\, sex\right) && \text{for } (r,s) \in \{(1,2),(2,3),(3,4)\} \\ q_{rs}(t) &= \exp\left(\beta_B + \xi_B t\right) && \text{for } (r,s) \in \{(2,1),(3,2),(4,3)\} \\ q_{rs}(t) &= \exp\left(\beta_D + \xi_D t + \gamma_D sex\right) && \text{for } (r,s) \in \{(1,5),(2,5),(3,5),(4,5)\}, \end{aligned}$$

where sex is coded as before by 0/1 for women/men. The subscripting by the letters is F for forward transitions, B for backward transitions, and D for transitions into the dead state. This is Model I.

For the frailty models, the individuals are considered as the clusters and index i is used to denote the individual-specific frailties. Estimation is by marginal likelihood, where Gauss–Hermite quadrature (with 15 nodes) is used for the models with normal frailty distributions, and the trapezoidal rule (with 15 nodes) is used for the model with the gamma distribution. The maximisation of the marginal likelihood function is undertaken using the general-purpose optimiser `optim` in R.

In a five-state model such as the current one, there are many ways in which a frailty model can be defined. In practice, one has to take into account the information available in the data and the computational challenges when fitting extended frailty models. In the current example, we assume that interest is focussed on cognitive decline over time; that is, progressive movement through the states 1, 2, 3, and 4, with special attention to entry into state 4. Information on backward transitions is limited and for this reason frailties will not be included for the hazards of these transitions.

Model II is defined by including a frailty term for transition $3 \to 4$ using a log-normal frailty. The model is the same as Model I except for the specification of $q_{34}(t|i)$, which is conditional on cluster i, $i = 1, ..., N$. The specification is given by

$$\begin{aligned} q_{34}(t|i) &= \exp(\beta_{34} + B_{i.34} + \xi_F t + \gamma_F\, sex) \\ B_{i.34} &\sim N(0, \sigma_{34}^2). \end{aligned}$$

Model III is similar to Model II, but the frailty for transition $3 \to 4$ is defined

Table 5.1 *Log-likelihood values at the maximum for five-state models for the ELSA data on remembering words.*

Model		$-2\log(L_{\max})$
I	Fixed effects only	7776.7
Univariate frailty for $3 \to 4$		
II	Log-normal	7775.0
III	Gamma	7773.1
Univariate frailty for $2 \to 3$		
II	Log-normal	7767.3
III	Gamma	7765.4
Bivariate frailty for $2 \to 3$ and $3 \to 4$		
IV	Log-normal frailty	7762.9
Mover/stayer models		
V	Log-normal frailty	7738.9
VI	Log-normal frailty	7727.8

using the gamma distribution:

$$\begin{aligned} q_{34}(t|i) &= Z_{i.34}\exp(\beta_{34} + \xi_F t + \gamma_F\, sex) \\ Z_{i.34} &\sim Gamma(\theta). \end{aligned}$$

Table 5.1 presents the results for Models I, II, and III. Adding a frailty as a random intercept to the hazard model for transition $3 \to 4$ does not lead to a substantial improvement with respect to the log-likelihood. There is, however, some improvement if Models II and III are specified by a frailty term for transition $2 \to 3$; see also Table 5.1. The latter is the reason to define Model IV, which includes frailty terms for $2 \to 3$ and $3 \to 4$. Using the bivariate normal distribution, Model IV is given by

$$\begin{aligned} q_{23}(t|i) &= \exp(\beta_{24} + B_{i.23} + \xi_F t + \gamma_F\, sex) \\ q_{34}(t|i) &= \exp(\beta_{34} + B_{i.34} + \xi_F t + \gamma_F\, sex) \\ (B_{i.23}, B_{i.34}) &\sim N(\mathbf{0}, \mathbf{\Sigma}). \end{aligned}$$

This model is estimated by marginal likelihood where the bivariate

Gauss–Hermite quadrature is implemented using univariate normal distributions; see end of Section 5.3. The two univariate quadratures are defined using 15 nodes, which is computationally intensive as it implies evaluating the integrand 15×15 times for each individual contribution to the likelihood. The covariance 2×2 matrix $\boldsymbol{\Sigma}$ is not estimated explicitly, but can be derived from the estimated standard deviations σ_{23} and σ_{34}, and the correlation ρ. The estimates for Model IV are $\widehat{\sigma}_{23} = 0.712$ (0.142), $\widehat{\sigma}_{34} = 0.807$ (0.251), and $\widehat{\rho} = -0.293$ (0.692), where estimated standard errors are in parentheses.

The Monte Carlo approximation of the bivariate integral for Model IV is implemented by drawing repeatedly 15 samples from the bivariate frailty distribution. This approximation is faster than the quadrature with resulting point estimates being very similar. The disadvantage is that there is no proper convergence with the Monte Carlo approximation if the number of nodes does not change. For the current model, point estimates are close to the estimates derived from marginal likelihood estimation. For example, the point estimates for the covariance are $\widehat{\sigma}_{23} = 0.759$, $\widehat{\sigma}_{34} = 0.714$, and $\widehat{\rho} = -0.338$. Although these values are not the same as the marginal likelihood estimates, taken into account the uncertainty as estimated by the marginal likelihood, the inference is similar.

What may work well in practice is to start with the fast Monte Carlo approximation, and use quadrature when close to the maximum of the likelihood.

The next step is to define *mover/stayer models*, where the frailty terms allow the distinction between individuals who move quickly through the states (the movers) and individuals who tend not to move so quickly (the stayers). Consider Model V defined by

$$\begin{aligned}
q_{rs}(t|i) &= \exp\left(\beta_{rs.0} + B_i + \xi_F t + \gamma_F \, sex\right) \quad \text{for } (r,s) \in \{(1,2),(2,3),(3,4)\} \\
q_{rs}(t|i) &= \exp\left(\beta_B + \xi_B t\right) \quad \text{for } (r,s) \in \{(2,1),(3,2),(4,3)\} \\
q_{rs}(t|i) &= \exp\left(\beta_D + B_i + \xi_D t + \gamma_D sex\right) \\
& \qquad \text{for } (r,s) \in \{(1,5),(2,5),(3,5),(4,5)\} \\
B_i &\sim N(0, \sigma^2).
\end{aligned}$$

An individual i with a large $B_i > 0$ is a mover; an individual i with a small $B_i < 0$ is a stayer. Given the current multi-state process, the stayers are individuals whose cognitive performance and survival are better than average when controlling for age and gender.

Model VI is defined by dropping random-effect B_i from the hazard submodel for death in Model V. This implies that Model VI only considers the mover/stayer distinction for the progression through the living states.

Table 5.2 *Parameter estimates for the five-state mover/stayer Model VI for the ELSA data on cognitive function. Estimated standard errors in parentheses. Time scale t is age in years minus 49.*

Intercept			t			*sex*		
$\beta_{12.0}$	0.162	(0.336)	ξ_F	0.053	(0.011)	γ_F	0.351	(0.099)
β_D	−6.009	(0.246)	ξ_D	0.101	(0.008)	γ_D	0.645	(0.142)
β_B	−0.133	(0.257)	ξ_B	−0.026	(0.010)			
$\beta_{23.0}$	−1.489	(0.259)				σ	0.953	(0.116)
$\beta_{34.0}$	−3.490	(0.244)						

Table 5.1 shows that Model VI performs better than Model V. Model V assumes that one frailty distribution is capable of capturing differences in survival *and* change in cognitive function. Apparently, this is not correct for the ELSA data. The process of survival is structurally different from change in cognitive function, which is illustrated by the better fit of Model VI.

With 10 transitions in the current multi-state survival model, there are a large number of frailty models that are potentially of interest. The examples above can be building blocks for extended modelling.

With Model VI the best model so far, we end this section with additional statistical inference. For the model parameters, Table 5.2 presents point estimates and estimated standard errors. The errors are based upon the Hessian evaluated at the maximum of the log-likelihood, where the delta method is used for the transformation $\sigma = \exp(\theta)$ with $\theta \in \mathbb{R}$.

To investigate the performance of the Monte Carlo EM algorithm in Section 5.4, this algorithm was implemented with number of Monte Carlo simulations $S = 5$ in the E-step, and with total number of EM iterations $V = 400$. The large value for V was chosen to monitor stability of the estimation. The starting values for the EM are the point estimates for the fixed-effects model derived with the scoring algorithm, and 1.5 for the standard deviation σ of the distribution of the random effects. Visual inspection of the trend in the parameter estimates shows that $V = 100$ would have sufficed in this application. Changing the starting value the standard deviation σ to 0.5 yields remarkable similar point estimates, but stability in the estimation was only reached after about 200 EM-iterations.

Point estimates as provided by the Monte Carlo EM are similar to the point estimates in Table 5.2. There are some differences, but for practical inference the result is the same. For the effects of age the estimates are $\widehat{\xi}_F =$

0.071, $\widehat{\xi}_D = 0.103$, and $\widehat{\xi}_B = -0.016$. The standard deviation is estimated at $\widehat{\sigma} = 0.982$, which is very close to the corresponding value in Table 5.2.

Returning to Model VI and the marginal likelihood estimates, for the estimation of individual random effects, maximum *a posteriori* (MAP) estimation is undertaken. Let $\boldsymbol{\theta}$ denote the vector with all the model parameters, and let $\mathbf{y}_i = (y_1, ..., y_{J_i})$ be the vector with the data for individual i. The density of frailty value b_i evaluated at the maximum likelihood estimate (MLE) of $\boldsymbol{\theta}$ is given proportionally by

$$f(b_i|\mathbf{y}_i;\boldsymbol{\theta}=\widehat{\boldsymbol{\theta}}) \quad \propto \quad P(\mathbf{y}_i|b_i;\boldsymbol{\theta}=\widehat{\boldsymbol{\theta}})f(b_i|\boldsymbol{\theta}=\widehat{\boldsymbol{\theta}}). \tag{5.8}$$

The first term at the right-hand side is the conditional likelihood contribution to the likelihood for individual i, and the second term is the density of the estimated frailty distribution, which is normal with mean zero and standard deviation equal to estimated σ. For each i, frailty value b_i can thus be estimated by maximising (5.8) using a general-purpose optimiser.

For the current example, the estimated b_i, $i = 1, ..., N$, are ordered from small to large. The observed trajectories of the ten individuals with estimated frailty in the lower 1% percentile are depicted in Figure 5.1. Compared to the age-specific mean for the number of words recalled (grey curve in Figure 5.1), these 10 individuals perform very well on the test for cognitive function. Likewise the 10 individuals with estimated frailty in the 1% upper percentile are depicted. The outlying behaviour of these two groups in the data is reflected in the estimated frailties in Model VI.

5.6 Non-parametric frailty distribution

The multi-state frailty models in Section 5.2 are defined using parametric distributions for the frailties. The current section will discuss models with non-parametric frailty distributions.

Non-parametric maximum likelihood (NPML) estimation of generalised linear mixed models is discussed in Aitkin (1999), Molenberghs and Verbeke (2005), Muthén and Asparouhov (2009), and Einbeck and Hinde (2009). The method is based upon finite mixtures of distributions. Mathematically, the same approach is used in latent-class analysis; see, for example, Vermunt and Magidson (2002).

By adopting the non-parametric approach, the assumption of normality for the random effects is avoided. The specification of the distribution of the random effects does not always have an impact on the estimation of the parameters of interest, but there are examples where the normality assumption leads to bias (Muthén and Asparouhov, 2009). The main advantage of the

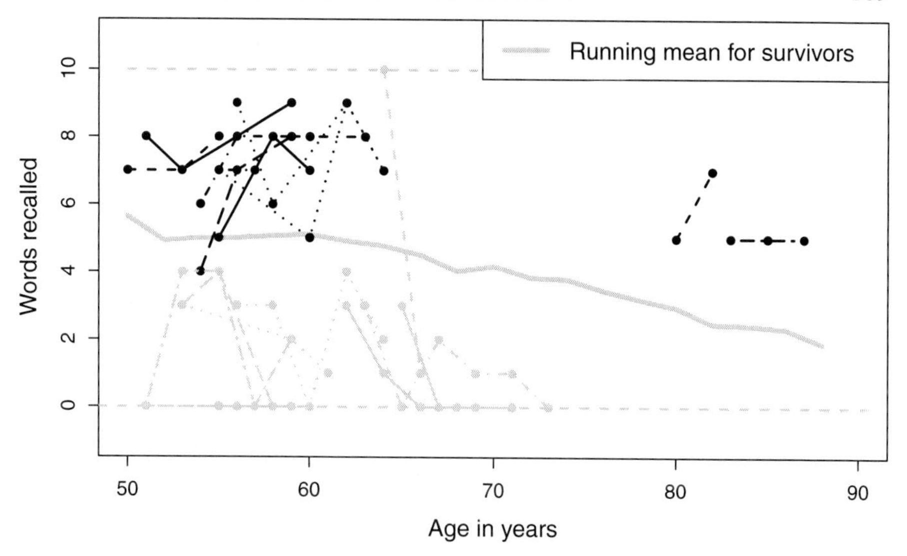

Figure 5.1 *Movers and stayers in the ELSA data according to frailty Model VI. Black trajectories for stayers (estimated frailty in lower 1% percentile), grey for movers (estimated frailty in upper 1% percentile).*

non-parametric approach is that it works well when the effects are normally distributed *and* when they are not.

Consider the presence of classes $\mathcal{C}_k$, $k = 1,...,K$, with the probability of class-membership $P(c \in \mathcal{C}_k)$ for cluster c unknown. The law of total probability for discrete random variables is used to define the likelihood function. If $\mathbf{y} = (y_1,...,y_J)$ are the longitudinally observed states of individual i in cluster c, then the likelihood function contribution of i is

$$\begin{aligned} L_i(\boldsymbol{\theta}|c,\mathbf{y},\mathbf{x}) &= P(Y_J = y_J,...,Y_2 = y_2|Y_1 = y_1, c, \boldsymbol{\theta}, \mathbf{x}) \\ &= \sum_{k=1}^{K} P(Y_J = y_J,...,Y_2 = y_2|Y_1 = y_1, \boldsymbol{\theta}, \mathbf{x}, c \in \mathcal{C}_k) P(c \in \mathcal{C}_k), \end{aligned}$$

where K is the number of classes. If the cluster is defined by the individual, then $c = i$. Denote $\pi_{ck} = P(c \in \mathcal{C}_k)$, and let $\boldsymbol{\theta}_k$ be the part of $\boldsymbol{\theta}$ that pertains to class $\mathcal{C}_k$, $k = 1,..,K$. With notation $\boldsymbol{\theta}_0$ for the model parameters which are not class-specific, we write $\boldsymbol{\theta} = (\boldsymbol{\theta}_0, \boldsymbol{\theta}_1,...,\boldsymbol{\theta}_K)$. The likelihood contribution is thus given by

$$L_i(\boldsymbol{\theta}|c,\mathbf{y},\mathbf{x}) = \sum_{k=1}^{K} P(Y_J = y_J,...,Y_2 = y_2|Y_1 = y_1, \boldsymbol{\theta}_0, \boldsymbol{\theta}_k, \mathbf{x})\, \pi_{ck}\,. \quad (5.9)$$

The class-specific parameters $\boldsymbol{\theta}_k$ are called the *mass points*, and the weights π_{ck} are called the *masses*. A straightforward simplification is to assume that the masses are cluster-independent; that is, $\pi_{ck} = \pi_k$, for $k = 1, ..., K$.

Combining contributions (5.9) for all N individuals, the likelihood function is given by $L = \prod_{i=1}^{N} L_i(\boldsymbol{\theta}|\mathbf{y}, \mathbf{x})$.

The number of classes K is an unknown model parameter. It is not straightforward to add K as a free parameter in the maximum likelihood estimation. Instead, we will conduct the likelihood estimation conditional on a specified K, and then compare models for various choices of K. This approach is not without problems. Note that models with different choices for K are not nested. The standard likelihood ratio test for model comparison cannot be applied to decide upon K. This is a more general problem of determining the distribution for the likelihood ratio test statistic in mixture models; see, for example, Aitkin et al. (2009).

Even though NPML has the same mathematical structure as used in latent class analysis, as an approach to model non-parametric random effects, NPML is different. In latent class analysis, the aim is to discover an interpretable classification in the data. The aim of NPML is to take into account unobserved heterogeneity—the classes are not interpreted necessarily. For this reason, an NMPL class with a very low mass is not a problem: that class just captures some undefined outlier behaviour. As a consequence, number of classes K in NMPL is not important with respect to interpretation (bar the obvious exception $K = 1$ versus $K > 1$).

Comparing NPML with parametric models for frailty distribution, the disadvantages are clear: the inference for K is not well-founded in theory and the number of parameters rises fast with increasing K. The advantages are that NPML can take into account non-normal distributions, and that there are no integrals in the likelihood function. If general-purpose optimisers are used for maximisation, then NPML models with small K are computationally less intensive than models with parametric frailties.

For generalised linear models, the EM algorithm has been proposed for maximum likelihood estimation; see, for example, Aitkin (1999). It is also possible to use general-purpose optimisers; see the non-linear models in Van den Hout et al. (2013).

The estimation of the masses π_k, $k = 1, ...K$, with restriction $\sum_{k=1}^{K} \pi_k = 1$ can be undertaken by using transformation such that parameters to be optimised are unrestricted in $\mathbb{R}$. As an example, consider the case $K = 3$. The three masses are estimated by estimating the two independent parameters

$\psi_1, \psi_2 \in \mathbb{R}$ such that

$$\pi_1 = \frac{1}{1+\exp(\psi_1)+\exp(\psi_2)}$$

$$\pi_2 = \frac{\exp(\psi_1)}{1+\exp(\psi_1)+\exp(\psi_2)} \qquad \pi_3 = \frac{\exp(\psi_2)}{1+\exp(\psi_1)+\exp(\psi_2)}.$$

This transformation, which works for any choice of K, is also used in multinomial logit models. And—as with the logit models—this transformation can be extended to regression models, where linear predictors can include cluster-specific covariate effects.

As commented by several authors, using NPML often leads to local maxima of the likelihood function; see, for example, Lindsay (1995, Section 1.5) and Aitkin et al. (2009, Section 7.4). This is an extra motivation to investigated sensitivity with respect to starting values for the maximisation. There is of course also an identifiability issue with the classes in the sense that the order of the classes $C_1,..,C_K$ is not identified from the data unless an order of the masses according to size is enforced.

Interestingly, when there are no restrictions on the number of classes and the NMPL likelihood function is viewed as a function of a completely unknown mixing distribution, then the solution has good properties. This is discussed in depth in Lindsay (1995). Further computational methods for estimating the mixing distribution can be found in Wang (2010).

5.7 Example: frailty in ELSA (continued)

This section illustrates using NPML models for multi-state survival processes. The application is a follow-up to Section 5.4 which explored parametric frailty models for the five-state process defined for the ELSA data on cognitive function and survival. As a sequence to the results in Section 5.4, the discussion in the current section will be restricted to models with univariate frailties and associated mover/stayer models.

In Section 5.4, fixed-effects Model I is introduced for the five-state survival model for the ELSA data. Model II is defined by extending Model I by including an individual-specific frailty term for the transition $3 \rightarrow 4$ using a log-normal frailty distribution. Define the NPML version of Model II as Model $\text{II}_{K=K}$, which is like Model II except for the definition of the class-specific hazard for the transition $3 \rightarrow 4$. For class C_k we define

$$q_{34}(t|k) = \exp\left(\beta_{k.34} + \xi_F t + \gamma_F sex\right),$$

where $\beta_{k.34}$, for $k = 1...,K$, are the mass points which together with the masses

$\pi_1, \ldots, \pi_K$ define the non-parametric distribution of the frailty for transition $3 \rightarrow 4$. We investigate models for $K = 2$ and for $K = 3$ with notation $\mathrm{II}_{K=2}$ and $\mathrm{II}_{K=3}$, respectively.

The general-purpose optimiser `optim` in R is used to maximise the likelihood function (5.9). An order of the estimated masses π_k is not enforced, so the solution of the maximisation is not unique with respect to the indexing of the classes. The delta method in Section 4.5 is applied to derive standard errors for estimated π_k after maximisation.

In applications of NPML and latent-class analysis, the Bayesian information criterion (BIC) is often used to choose K; see also Muthén and Asparouhov (2009). For the current analysis, we will report $-2\log(L_{\max})$, BIC, and Akaike's information criterion (AIC). In BIC definition (4.11), we use n equal to the number of individuals in the data. Interpretation of differences will be heuristic as the assumptions for these criteria do not hold for models estimated by NPML.

Table 5.3 presents the information criteria for the fixed-effects model and for a series of NPML models. The fixed-effects model is the same as the one in Table 5.1. Using NPML to include either a frailty for $3 \rightarrow 4$ or a frailty for $2 \rightarrow 3$ does not seem worthwhile according to the BIC. However, using NPML to define mover/stayer models for $K = 2$, 3, and 4 leads to substantial reduction in all three criteria.

Mover/stayer Model $\mathrm{VI}_{K=K}$ includes a frailty term for the forward transitions through the living states. The model is defined using a sum-to-zero constraint for the mass points, where the mass points are additive effects in the log-linear predictor for the forward transition through the living states. Model $\mathrm{VI}_{K=K}$ is thus given by

$$\begin{aligned} q_{rs}(t|k) &= \exp\left(\beta_{rs.0} + b_k + \xi_F t + \gamma_F\, sex\right) \quad \text{for } (r,s) \in \{(1,2),(2,3),(3,4)\} \\ q_{rs}(t|k) &= \exp\left(\beta_B + \xi_B t\right) \quad \text{for } (r,s) \in \{(2,1),(3,2),(4,3)\} \\ q_{rs}(t|k) &= \exp\left(\beta_D + \xi_D t + \gamma_D\, sex\right) \\ &\quad \text{for } (r,s) \in \{(1,5),(2,5),(3,5),(4,5)\}, \end{aligned}$$

with constraint $\sum_{k=1}^{K} b_k = 0$.

According to the BIC, the model with $K = 2$ performs best. This implies that for the data at hand, two classes can be distinguished with respect to latent frailty terms b_k. This two-class model links up nicely with the mover/stayer interpretation.

Table 5.4 presents parameter estimates for Model $\mathrm{VI}_{K=2}$. Given $K = 2$ and the sum-to-zero constraint for the mass points, there is only one independently estimated mass point in Model $\mathrm{VI}_{K=2}$. In the maximum likelihood estimation,

Table 5.3 *Five-state models for the ELSA data on remembering words. Information criteria for the fixed-effects model and frailty models estimated by non-parametric maximum likelihood (NPML). BIC penalty defined using the number of individuals ($N = 1000$) as the sample size.*

Model	Number of parameters	$-2\log(L_{\max})$	BIC	AIC
Fixed effects only				
I	10	7776.7	7845.7	7796.7
Frailty for $3 \to 4$				
$\text{II}_{K=2}$	12	7774.1	7856.9	7798.1
$\text{II}_{K=3}$	14	7774.1	7870.8	7802.1
Frailty for $2 \to 3$				
$\text{II}_{K=2}$	12	7767.9	7850.8	7791.9
$\text{II}_{K=3}$	14	7767.4	7864.1	7795.4
Mover/stayer models				
$\text{VI}_{K=2}$	12	7724.7	7807.7	7748.8
$\text{VI}_{K=3}$	14	7716.0	7812.7	7744.0
$\text{VI}_{K=4}$	16	7713.4	7823.9	7745.4

Table 5.4 *For the ELSA data on cognitive function, parameter estimates for the five-state mover/stayer non-parametric maximum likelihood (NPML) Model $VI_{K=2}$. Estimated standard errors in parentheses. Time scale t is age in years minus 49.*

Intercept			t			*sex*		
$\beta_{12.0}$	−0.744	(0.198)	ξ_F	0.053	(0.008)	γ_F	0.270	(0.080)
β_D	−6.015	(0.247)	ξ_D	0.104	(0.008)	γ_D	0.533	(0.141)
β_B	−0.430	(0.134)	ξ_B	−0.020	(0.007)			
$\beta_{23.0}$	−2.220	(0.192)						
$\beta_{34.0}$	−3.797	(0.205)	*Class C_1 parameters:*			b_1	−0.823	(0.092)
						π_1	0.211	(0.050)

this mass point is b_1, which is estimated at -0.823 with estimated standard error 0.092. The mass π_1 is estimated at 0.211 with estimated standard error 0.05. This means that the probability to be in class C_1 is estimated at 0.211,

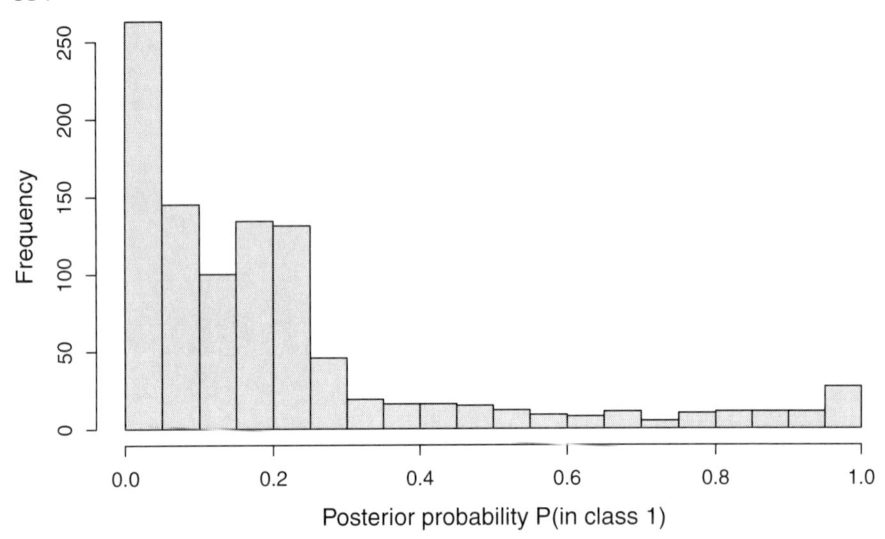

Figure 5.2 *For the ELSA data and the mover/stayer Model $VI_{K=2}$, histogram of the estimated $N = 1000$ posterior probabilities for being in the stayer-class C_1.*

and that this class represents the stayers, as the class-specific effect on the log-linear predictor for the hazard is negative implying that individuals in class C_1 are less likely to move to the next living state than individuals in class C_2.

The parameter estimates in Table 5.4 can be compared to the estimates of the parametric frailty model in Table 5.2. Note that the ξ-parameters for the effect of age are very similar, but that the effects of gender are attenuated in the NMPL model.

It is interesting to look at the individual class-membership probabilities derived from the estimated model. These are called posterior probabilities and are derived using Bayes' theorem. Conditional on individual data $\mathbf{y} = \mathbf{y}_i$ and vector $\widehat{\boldsymbol{\theta}}$ with the estimates of the model parameters, it follows that

$$P(i \in C_k | \mathbf{Y} = \mathbf{y}; \boldsymbol{\theta} = \widehat{\boldsymbol{\theta}}) = \frac{P(Y_J = y_j, ..., Y_2 = y_2 | Y_1 = y_1, i \in C_k; \boldsymbol{\theta} = \widehat{\boldsymbol{\theta}}) P(i \in C_k | \boldsymbol{\theta} = \widehat{\boldsymbol{\theta}})}{\sum_{\ell=1}^{K} P(Y_J = y_j, ..., Y_2 = y_2 | Y_1 = y_1, i \in C_\ell; \boldsymbol{\theta} = \widehat{\boldsymbol{\theta}}) P(i \in C_\ell | \boldsymbol{\theta} = \widehat{\boldsymbol{\theta}})}.$$

This probability can be derived from the estimated model since $P(i \in C_k | \boldsymbol{\theta} = \widehat{\boldsymbol{\theta}})$ is estimated by mass $\widehat{\pi}_k$, and $P(Y_J = y_j, ..., Y_2 = y_2 | Y_1 = y_1, i \in C_k; \boldsymbol{\theta} = \widehat{\boldsymbol{\theta}})$ is the likelihood contribution for individual i conditional on class membership $i \in C_k$.

For Model $VI_{K=2}$, Figure 5.2 presents the histogram of the estimated $N = 1000$ posterior probabilities for being in the stayer-class C_1. Given that the marginal mass π_1 is estimated at 0.211, it is to be expected that the histogram shows positive skew.

The posterior probabilities can be further explored. One might wonder whether there is an association between posterior probabilities and the state at the start of the follow-up. This association was investigated by a one-way analysis of variance (ANOVA) with the logarithm of the posterior probability as the response. Baseline state is included as a factor with four levels.

The fit of the ANOVA was checked by residual diagnostics, which did not flag serious problems. Choosing a significance level of 5%, the ANOVA table shows that the null-hypothesis of no factor effect can be rejected: the p-value for the standard F-test is less than 0.01.

Means of the response can be defined for each of the four factor levels. Figure 5.3 presents the result of Tukey's range test, which investigates which means are significantly different from each other taking into account the multiple testing. With four factor levels, there are six possible mean comparisons. Using again the significance level of 5%, Figure 5.3 shows that there are significant differences in mean response when comparing baseline state 1 with the other states. The mean of the posterior probabilities for those individuals who are observed in state 1 at baseline are significantly larger than the mean of the posterior probabilities for those individuals who are observed in one of the other states at baseline. Going back to the interpretation of the latent classes, this means that individuals who are observed in state 1 at baseline are more likely to be classified in the stayer-class C_1 than individuals who are observed in state 2, 3, or 4 at baseline.

At baseline, there are 99 individuals observed in state 1. If the risk of moving to the next living state as described by the fixed effect parameters is mainly determined by the other $N - 99 = 901$ individuals, then the frailty model can be understood in the sense that it classifies the 99 individuals who are observed in state 1 to be in the stayer-class C_1: these individuals have a lower risk of moving compared to the other individuals. This result is as expected. Being observed in state 1 at baseline implies good cognitive function, which is often associated with good health and thus with a lower risk of cognitive decline.

Exploration of posterior probabilities may also lead to unexpected results. One might wonder whether there is an association between posterior probabilities and the number of times individuals are observed in the data. This association was also investigated by an ANOVA with the logarithm of the posterior probability as the response. The number of observations was included as a factor, where death was counted as an observation and the five factor levels are defined by 2, 3, 4, 5, and 6 observations. The outcome of this

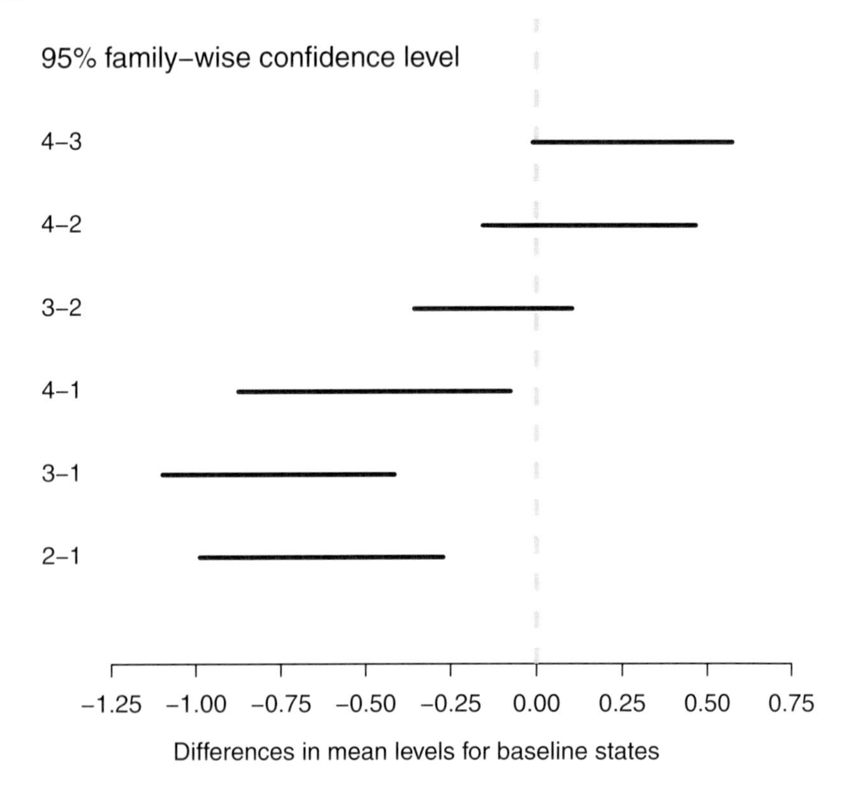

Figure 5.3 *For the ELSA data and the mover/stayer Model* $VI_{K=2}$, *post-hoc ANOVA analysis: Tukey's range test for the effect of baseline state (1, 2, 3, or 4) on the logarithm of the posterior probability for being in the stayer-class* C_1.

ANOVA was that individuals who are observed more often are more likely to be classified as movers.

This outcome was unexpected in the sense that if individuals are movers, then they are typically less healthy and thus more likely to die during the follow-up, which results in a lower number of observation times when compared to the healthy individuals. Another reason that this outcome is unexpected is that individuals who are lost to follow-up in ELSA tend to be less wealthy, more ill, and less educated than those who did not drop out (Steptoe et al., 2013). If they are more ill, one would expect that they move more quickly through the living states.

The association between the frailty terms and observable features of the data needs more research. A tentative explanation of the second ANOVA is that the ELSA follow-up is not long enough to induce an effect of lost to follow-up.

Although there is no indication of doctor's care in ELSA, this observation scheme may play a role in other studies; see Section 1.2. In the presence of doctor's care, individuals with poorer health are observed more often and this induces a dependency between state occupancy and the observation process. With an NMPL model such a dependency may show in an exploration of the posterior probabilities.

In general, if post-hoc analyses do not yield any significant association between the frailty terms and observable features of the data, then the posterior probabilities are more likely to be indicative of latent traits in the individuals in the data. This might justify additional analyses; for instance, using the posterior probability as an explanatory variable in other models.

Chapter 6

Bayesian Inference for Multi-State Survival Models

6.1 Introduction

This chapter introduces methods for Bayesian inference for continuous-time multi-state survival models. After a preamble on the Bayesian framework, methods for inference will be presented and illustrated. The main aim of this chapter is to show how basic methods for Bayesian inference can be used to analyse multi-state survival data.

There is a wide range of introductory books on Bayesian inference. A good start is Gelman et al. (2004). Even though this book does not discuss continuous-time multi-state models, the approach and methods that are presented are relevant. Another source is Lunn et al. (2013), which includes an application using a discrete-time Markov model. Specialist books on topics in statistics can also have good introductions to Bayesian inference; Johnson and Albert (1999) and Fox (2010) are examples. Johnson and Albert (1999) is also the inspiration for the graphical illustration in Section 6.3.

Specifically for multi-state models, methods for Bayesian inference can be found in, for example, Sharples (1993), Welton and Ades (2005), Pan et al. (2007), Kneib and Hennerfeind (2008), and Van den Hout et al. (2015).

Say scalar parameter θ is of interest. In a frequentist framework, parameter θ is considered to be a fixed but unknown value. Frequentist inference with respect to statistical uncertainty is based upon frequency properties of the estimation under hypothetical repetition of data collection and data analysis. In a Bayesian framework, parameter θ is considered to be a random variable and of interest is its distribution given the data. A Bayesian does not try to estimate an unknown value, but an unknown distribution. This distribution is called the *posterior distribution* and will be denoted by $p(\theta|\mathbf{y})$, where $\mathbf{y}$ represents the data.

In line with many texts on Bayesian inference, notation is simplified. General notation $p()$ is used for probability density functions and probability mass functions. And data are only referred to by small letters: instead of $p(\mathbf{Y} = \mathbf{y})$, we simple write $p(\mathbf{y})$.

It follows that

$$p(\theta|\mathbf{y}) = \frac{p(\theta, \mathbf{y})}{p(\mathbf{y})} = \frac{p(\mathbf{y}|\theta)p(\theta)}{p(\mathbf{y})} \propto p(\mathbf{y}|\theta)p(\theta), \tag{6.1}$$

where $\propto$ stands for *proportional to*. If the aim is inference on the posterior mean of θ, or posterior uncertainty, then the marginal distribution $p(\mathbf{y})$ can be ignored. In (6.1), distribution $p(\mathbf{y}|\theta)$ is the likelihood function, and $p(\theta)$ is called the *prior distribution*.

The prior distribution contains the information about θ prior to the analysis of data $\mathbf{y}$. The relation $p(\theta|\mathbf{y}) \propto p(\mathbf{y}|\theta)p(\theta)$ reflects the idea that the prior distribution is updated by the information in the data with the posterior distribution as the result.

It is possible to specify vague prior distributions which are uninformative. For example, if θ is the height in meters of the human population of London, then $p(\theta) \sim U(0,3)$ is a vague prior distribution and the only information that is going to be relevant in data analysis is provided by the data. Another way to work with no prior information is to specify prior distributions that are not properly defined. In the above example, if we specify $p(\theta) \propto 1$, then the posterior distribution becomes $p(\mathbf{y}|\theta)$.

For certain types of Bayesian inference it is essential to specify the prior distribution properly. Using Bayes factors for model comparison is an example of this. The Bayes factor for models $\mathcal{M}_1$ and $\mathcal{M}_2$ is given by

$$B_{12} = \frac{p(\mathbf{y}|\mathcal{M}_1)}{p(\mathbf{y}|\mathcal{M}_2)},$$

where $p(\mathbf{y}|\mathcal{M}_i) = \int p(\mathbf{y}|\theta_i)p(\theta_i)d\theta_i$ is the marginal likelihood of the data under $\mathcal{M}_i$, for $i = 1,2$; see, for example, Bernardo and Smith (2000, Section 6.1.4). If the prior distribution is not a distribution, then the marginal likelihood is not properly defined and this invalidates the definition of the Bayes factor.

The basic method for Bayesian inference that will be used throughout this book is sampling from the posterior distribution. Once a sample is obtained, inference can start by looking at posterior means and posterior credible intervals. The latter are the Bayesian variants of the frequentist confidence intervals, but with a different interpretation: a 95%-credible interval for θ is an interval (θ_L, θ_U) such that $F(\theta_U) - F(\theta_L) = 0.95$ for F, the cumulative distribution function. For model comparison, we will use the deviance information criterion (DIC, Spiegelhalter et al., 2002), which will be introduced in Section 6.5.

The sampling from the posterior distribution will be undertaken by Markov chain Monte Carlo (MCMC) methods. Using MCMC can be seen as a third numerical method of estimating model parameters in addition to maximum likelihood estimation by either using a general-purpose optimiser (Section 4.5) or the scoring algorithm (Section 4.6).

For fixed-effects models with vague prior distributions, the MCMC can be seen as an exploration of the likelihood. If asymptotic properties of the maximum likelihood estimation hold, maximum-likelihood point estimates and confidence intervals should be similar to posterior means and posterior credible intervals. This will be illustrated. As a method of sampling, there is nothing Bayesian about MCMC. Nevertheless, defining vague prior distributions so that MCMC can be used to explore the likelihood function is a mix-up of concepts since using the MCMC in this setting implies that the model parameters have a distribution. If one is prepared to ignore this incongruity, then MCMC can also be a practical tool in a frequentist framework; see, for example, Tanner (1996).

One important reason to extend inference to the Bayesian framework is that this framework is very suitable for the frailty models introduced in Chapter 5. This will be illustrated in Section 6.4.

6.2 Gibbs sampler

MCMC is a general method that can be used to sample from the posterior distribution over the unknown parameters. Because the samples are drawn sequentially such that the distribution of the next sample depends on the current sample and not on earlier draws, the draws form a Markov chain (Gelman et al., 2004, Section 11.2). The aim of MCMC is to construct a Markov chain that converges to a unique stationary distribution that coincides with the target posterior distribution.

The MCMC method we use is a Gibbs sampler where each parameter is sampled conditional on the other parameters and the data (Geman and Geman, 1984). In case there is no closed form of the conditional probability distribution, Metropolis sampling is undertaken (Metropolis et al., 1953). This scheme is sometimes known as Metropolis-within-Gibbs although some authors dislike this term; see the discussion in Carlin and Louis (2009, Section 3.4.4).

The idea of sampling each parameter conditionally can be explained best with an example. Consider the case of the bivariate posterior distribution of (θ_1, θ_2). The conditional distributions are given by

$$\begin{aligned} p(\theta_1|\theta_2, \mathbf{y}) &\propto p(\mathbf{y}|\theta_1, \theta_2)p(\theta_1|\theta_2) \\ p(\theta_2|\theta_1, \mathbf{y}) &\propto p(\mathbf{y}|\theta_1, \theta_2)p(\theta_2|\theta_1). \end{aligned}$$

If we assume that the prior distributions of the θs are independent, then

$$\begin{aligned} p(\theta_1|\theta_2, \mathbf{y}) &\propto p(\mathbf{y}|\theta_1, \theta_2)p(\theta_1) \\ p(\theta_2|\theta_1, \mathbf{y}) &\propto p(\mathbf{y}|\theta_1, \theta_2)p(\theta_2), \end{aligned}$$

where $p(\mathbf{y}|\theta_1, \theta_2)$ is the likelihood function, and $p(\theta_1)$ and $p(\theta_2)$ are the prior distributions.

Geman and Geman (1984) show that if values are sampled iteratively from these two conditional distributions, then bivariate samples can be obtained from the posterior distribution $p(\theta_1, \theta_2|\mathbf{y})$. To obtain a proper sample from this distribution, the first values of the iteratively sampled parameter values are ignored in order to lose the dependency on the starting values. The set of ignored values is called the *burn-in*.

The Gibbs sampler is illustrated with the three-state Parkinson's disease data introduced in Section 3.7. The states are the dementia-free state 1, the dementia state 2, and the dead state. Exponential hazards for the progressive model are defined by

$$q_{rs}(t) = \exp(\beta_{rs}) \qquad \text{for} \qquad r,s \in \{(1,2),(1,3),(2,3)\}.$$

We start with the two-parameter model using the restriction $\beta_{12} = \beta_{23}$. Prior distributions for the two parameters are specified independently and are set to 1. Hence the posterior distribution is proportional to the likelihood function.

The Metropolis steps in the Gibbs sampler are as follows. At sample iteration $m+1$ with current values $\beta_{12}^{(m)}$ and $\beta_{13}^{(m)}$, a candidate value $\beta_{12}^{(c)}$ is drawn from the normal distribution $N(\beta_{12}^{(m)}, \sigma_{12}^2)$. The conditional distributions are given by

$$\begin{aligned} p_m &= p(\beta_{12}^{(m)}|\mathbf{y}, \beta_{13}^{(m)}) \propto p(\mathbf{y}|\beta_{12}^{(m)}, \beta_{13}^{(m)}) \\ p_c &= p(\beta_{12}^{(c)}|\mathbf{y}, \beta_{13}^{(m)}) \propto p(\mathbf{y}|\beta_{12}^{(c)}, \beta_{13}^{(m)}), \end{aligned}$$

where $\mathbf{y}$ are the multi-state data. Note that the distributions at the right-hand

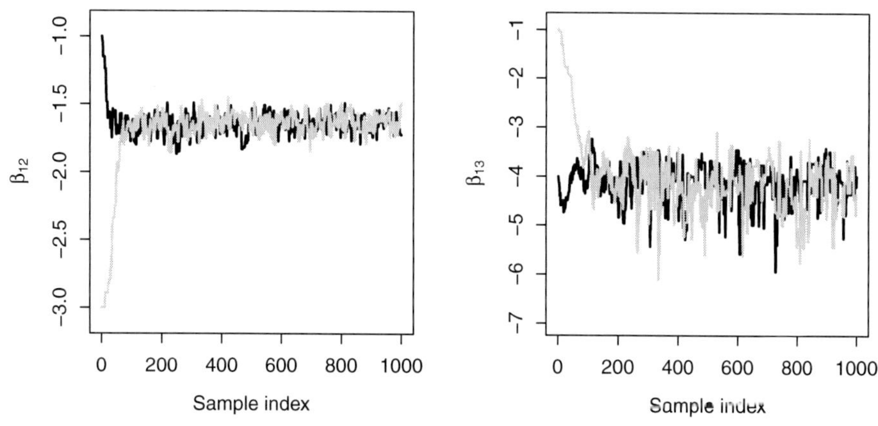

Figure 6.1 *Markov chain Monte Carlo sampling for the two-parameter model for the Parkinson's disease data. Using two chains (black and grey lines).*

side are the likelihood functions due to the specific choice of improper prior distributions. Next the probability

$$p = \min\left\{\frac{p_c}{p_m}, 1\right\}$$

is compared to a draw u from $U(0,1)$. If $u < p$, then $\beta_{12}^{(m+1)} = \beta_{12}^{(c)}$ else $\beta_{12}^{(m+1)} = \beta_{12}^{(m)}$. Analogously, the next value of β_{13} is drawn using the conditional distributions specified conditional on $\beta_{12}^{(m+1)}$.

The tuning parameters σ_{12} and σ_{13} are adapted during the sampling of the burn-in. We aim to keep the rate within the interval (0.30, 0.60) by the tuning of σ_{12} and σ_{13}. Basically, if too many values are accepted the corresponding tuning parameter is increased; if too few are accepted the parameter is decreased.

In the applications, two chains of 1000 sampled bivariate values are used. Starting values for $(\beta_{12}^{(0)}, \beta_{13}^{(0)})$ for the two chains are $\{(-3,-1),(-1,-4)\}$. Figure 6.1 shows the two chains of sampled values for each of the two β-parameters. The inference using MCMC does not pose any problems. The chains are well-behaved and a burn-in of 200 sampled values seem sufficient.

Figure 6.2 shows the two-dimensional estimation. The graph at the left-hand side shows the trajectories of the two chains toward the area with the highest posterior density. The two densities at the right-hand side are the univariate densities estimated from the two chains after discarding the first 200 sampled values in each chain as the burn-in.

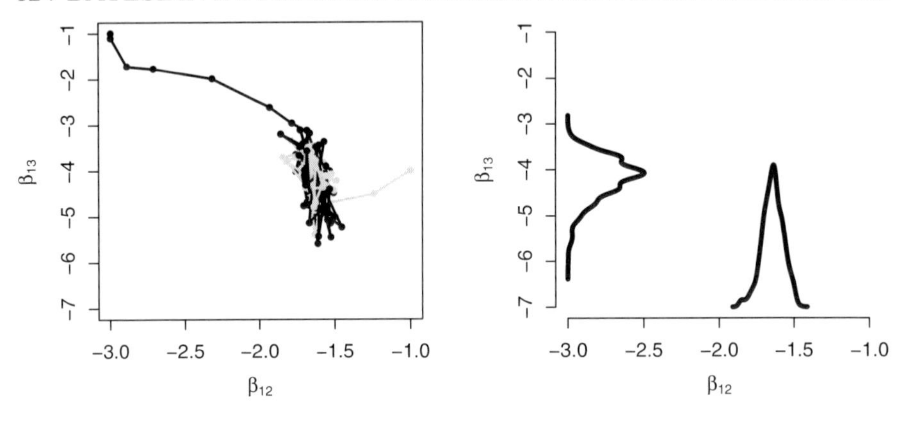

Figure 6.2 *At left-hand side, every 10th value in the bivariate Markov chain Monte Carlo sampling for the two-parameter model for the Parkinson's disease data. Black and grey lines for the two chains. At the right-hand side, kernel density estimates of posterior densities (proportional) after discarding the burn-in.*

Table 6.1 *Posterior inference for restricted intercept-only model fitted to the Parkinson's data.*

	Mean	SD	2.5%-Quantile	Median	97.5%-Quantile
$\beta_{12} = \beta_{23}$	-1.642	0.070	-1.798	-1.640	-1.509
β_{13}	-4.189	0.462	-5.272	-4.115	-3.458

The posterior densities in Figure 6.2 are derived by kernel density estimation, where the bandwidth is chosen according to Silverman's rule of thumb (Silverman, 1986, p. 48, Equation 3.31). There are many ways to define a kernel density estimator, some of which will produce curves more smooth than the ones in Figure 6.2. For reason of consistency, Silverman's rule of thumb is used for all the inferences in this chapter.

Using the remaining 1600 sampled values after the burn-in, posterior inference regarding means, medians, standard deviations, and credible intervals can be derived. Results are presented in Table 6.1. The posterior means and the corresponding posterior medians are close in value. This can also be inferred from the right-hand side of Figure 6.2, which shows that the posterior distributions of the parameters are symmetrical.

Once the Gibbs sampler is implemented for two parameters, estimation for extended models is straightforward. The Gompertz baseline hazard specification of the three-state progressive model is given by

$$q_{rs}(t) = \exp(\beta_{rs} + \xi_{rs} t) \qquad \text{for} \qquad r,s \in \{(1,2),(1,3),(2,3)\}$$

Table 6.2 *Bayesian inference for baseline models fitted to the Parkinson's data. Posterior means with posterior standard deviations in parentheses.*

Exponential			Gompertz			Weibull		
β_{12}	−1.97	(0.10)	β_{12}	−4.73	(0.58)	β_{12}	−13.00	(1.94)
β_{13}	−3.96	(0.40)	β_{13}	−6.92	(2.07)	β_{13}	−9.65	(5.60)
β_{23}	−1.36	(0.08)	β_{23}	−3.11	(0.61)	β_{23}	−9.28	(2.68)
			ξ_{12}	0.07	(0.01)	τ_{12}	3.68	(0.49)
			ξ_{13}	0.08	(0.05)	τ_{13}	2.40	(1.33)
			ξ_{23}	0.04	(0.01)	τ_{23}	2.83	(0.64)

(see Section 3.2.3), and the Weibull baseline hazard specification

$$q_{rs}(t) = \tau_{rs} t^{\tau_{rs}-1} \exp(\beta_{rs}) \qquad \text{for} \qquad r,s \in \{(1,2),(1,3),(2,3)\}$$

(see Section 3.2.2). The estimation uses the piecewise-constant approximation as defined in Section 4.4. Table 6.1 presents the posterior inference. This inference can be compared with the maximum likelihood estimation in Chapter 3, Table 3.4. And although there are small differences between the results in these tables, when the standard deviations (Table 6.2) and the standard errors (Table 3.4) are taken into account, the statistical inference is quite similar. Note that when a posterior distribution is skewed, the posterior mean of the distribution is not the maximum of the distribution.

One of the advantages of using MCMC is that there is no need to use the delta method for parameters with a restricted parameter space. For the Weibull model, the restriction is $\tau_{rs} > 0$, for all relevant r and s. Both for using maximum likelihood estimation and using MCMC, it is easier to define $\tau_{rs} = \exp(\gamma_{rs})$ and estimate unrestricted γ_{rs}. Section 4.5 showed that the delta method can be used to derive standard errors for $\widehat{\tau}_{rs}$ when the likelihood is maximised over γ_{rs}. When MCMC is used, this is not necessary. Exponentiating all the values sampled from the posterior distribution of γ_{rs} provides a sample from the posterior distribution of τ_{rs}. Posterior means, standard deviations, and credible intervals for τ_{rs} are thus directly available.

The Gibbs sampler for the extended models in Table 6.2 was implemented with two chains each with 2×10^5 iterations. The first half of each chain was used as burn-in and thus not used for the posterior inference. There are no general rules for determining the number of iterations beforehand, although the more the better is a good start. A number of diagnostic tools are proposed in the literature to check whether (part of) an MCMC chain is stable and can be used for posterior inference; see, for example, Gelman et al. (2004) or the options in the R package CODA (Plummer et al., 2006).

One of the diagnostic tools is the *potential scale reduction factor* introduced by Gelman and Rubin (1992). This factor monitors MCMC output with two or more chains. Chains are considered stable when the chains have become independent of their starting values and the output from the separate chains is indistinguishable. The factor has a univariate and a multivariate variant. The benchmark value is 1, with values between 1 and 1.1 considered to be acceptable in most cases (Gelman et al., 2004).

For the posterior inference presented in Table 6.2, both the univariate and multivariate potential scale reduction factor was between 1 and 1.001 for the Gompertz model. For the Weibull, the maximum of the six univariate factors was 1.05, and the multivariate factor was 1.05. So according to the potential scale reduction factor, MCMC works well for both models and can be used for posterior inference.

Even if starting values are provided close to the mean of the posterior density, sampling values using MCMC will also include sampling from the support where the density is low. Consider the Gompertz specification $\exp(\beta_{rs} + \xi_{rs} t)$ for the baseline hazard for the transition $r \rightarrow s$ at age t. Large values of ξ_{rs} can lead to numerical problems in the computation of the matrix with the transition probabilities. One way to deal with this problem is to restrict the support of the posterior density by using an informative prior density. Another way is to implement ad-hoc solutions to prevent numerical overflow when running the MCMC. In both cases, inspection of the chains and the resulting posterior density are recommended to check whether the chosen solution has an undesirable effect on statistical inference.

6.3 Deviance information criterion (DIC)

To compare models in a Bayesian framework, we use the Deviance information criterion (DIC, Spiegelhalter et al., 2002). The DIC comparison is based on a trade-off between the fit of the data to the model and the complexity of the model. Models with smaller DIC are better supported by the data. The DIC is based on the deviance, which is specified by

$$D(\mathbf{y}, \boldsymbol{\theta}) = -2 \log p(\mathbf{y}|\boldsymbol{\theta}), \tag{6.2}$$

where $\boldsymbol{\theta}$ is vector with all model parameters. The DIC is given by

$$\text{DIC} = \widehat{D} + 2p_D, \tag{6.3}$$

where $\widehat{D} = D(\mathbf{y}, E(\boldsymbol{\theta}))$ is the *plug-in deviance* and p_D denotes the effective number of parameters. The *expected deviance* is denoted $\overline{D}$ and is used to define p_D as $\overline{D} - \widehat{D}$. The plug-in deviance is estimated by using the posterior means of the model parameters. The expected deviance is estimated by $M^{-1} \sum_{m=1}^{M} D(\mathbf{y}, \boldsymbol{\theta}^{(m)})$, with m indexing parameter values which are sampled from the posterior.

The plug-in deviance is not invariant to parameterisation and does not take into account the precision of the estimates. The expected deviance, however, is a function of the posterior of the model parameters and does account for the precision of the estimates (Plummer, 2008).

Although the DIC is widely used, it is not without problems; see, for example, Carlin and Louis (2009) and the discussion in the seminal paper Spiegelhalter et al. (2002). The DIC can give inappropriate results if there are highly non-normal posterior distributions of the parameters on which prior distributions have been placed (Lunn et al., 2009). Some caution when using the DIC is therefore recommended.

For fixed-effects models and vague prior distributions for the parameters, the DIC is approximately equal to Akaike's information criterion (AIC), which is presented in Section 4.7, and the estimated number of effective parameters should be close to the total number of parameters. For frailty models, the number of effective parameters is in-between the number of independent model parameters and the total number of parameters as defined by model parameters *and* the individual random effects.

To illustrate model selection using DIC, consider the four models discussed for the Parkinson's disease data in the previous section. The exponential model with only two parameters has DIC = 1406.0, and the number of effective parameters is estimated at $p_D = 1.7$. The exponential model with three parameters has DIC = 1383.7 with $p_D = 3.1$. As expected, the second model is better according to the DIC than the first. As a rough guideline for comparing DICs, a difference of more than 5 can be considered substantial, and a difference of less than 5 should be interpreted with care (Lunn et al., 2013, Section 8.6.4). The Gompertz model has DIC = 1340.0 ($p_D = 5.7$), and the Weibull has DIC = 1337.5 ($p_D = 5.6$). Although the Weibull model has a lower DIC, the difference with the DIC for the Gompertz model is not substantial.

To illustrate a problem with the DIC, this criterion cannot be used to compare the non-parametric maximum likelihood (NPML) models in Sections 5.6 and 5.7. This is the more general issue that the DIC cannot be used for finite mixture models (Lunn et al., 2013, Section 8.6.6).

6.4 Example: frailty in ELSA (continued)

Returning to the example with the data from the English Longitudinal Study of Ageing (ELSA), this section illustrates using MCMC to estimate a multi-state survival model with more than three states. Using the subsample with $N = 1000$ individuals as detailed in Section 4.8.1, Bayesian inference will be presented for the mover/stayer frailty models defined in Section 5.4. Instead of integrating out the random effects in the marginal likelihood, all the parameters (the fixed effects, the random effects, and the variance component) are sampled from their conditional posterior distributions.

The presentation will be limited to three models. For ease of reference, the definitions of the models are repeated. The five-state survival model for the recall-of-words data in ELSA defines 10 transitions; see Figure 4.2. The linear fixed-effects regressions for the hazards are given by

$$\begin{aligned} q_{rs}(t) &= \exp\left(\beta_{rs.0} + \xi_{rs} t + \gamma_F \, sex\right) \qquad \text{for } (r,s) \in \{(1,2),(2,3),(3,4)\} \\ q_{rs}(t) &= \exp\left(\beta_B + \xi_B t\right) \qquad \text{for } (r,s) \in \{(2,1),(3,2),(4,3)\} \\ q_{rs}(t) &= \exp\left(\beta_D + \xi_D t + \gamma_D sex\right) \\ & \qquad \text{for } (r,s) \in \{(1,5),(2,5),(3,5),(4,5)\}, \end{aligned}$$

where the time scale t is age in years minus 49, and $sex = 0/1$ for men/women.

The first mover/stayer model is defined by including an individual-specific frailty $b_i \sim N(0, \sigma^2)$ in the hazard models for forward transitions:

$$\begin{aligned} q_{rs}(t|i) &= \exp\left(\beta_{rs.0} + b_i + \xi_{rs} t + \gamma_F \, sex\right) \quad \text{for } (r,s) \in \{(1,2),(2,3),(3,4)\} \\ q_{rs}(t|i) &= \exp\left(\beta_B + \xi_B t\right) \qquad \text{for } (r,s) \in \{(2,1),(3,2),(4,3)\} \\ q_{rs}(t|i) &= \exp\left(\beta_D + b_i + \xi_D t + \gamma_D sex\right) \\ & \qquad \text{for } (r,s) \in \{(1,5),(2,5),(3,5),(4,5)\} \\ b_i &\sim N(0, \sigma^2). \end{aligned}$$

The second mover-stayer model is defined by including $b_i \sim N(0, \sigma^2)$ only in the log-linear regressions for the forward transitions through the living states; that is, for $r \to s$ with $(r,s) \in \{(1,2),(2,3),(3,4)\}$.

For each of the 10 fixed-effects parameters improper univariate prior densities are specified by $p(\cdot) \propto 1$. For the variance component σ^2, the vague prior density is $U(0,4)$. For each of three models, the MCMC consists of two chains with each 10,000 sample iterations. Of each chain, the first 4,000 sampled values are discarded as the burn-in. The sampling of the values in the MCMC from the burn-in onward is checked by visual inspection and by computing the potential scale reduction factors.

The fixed-effects model with 10 parameters has DIC = 7796, and $p_D =$ 9.8. As expected, the estimated number of effective parameters is close to the total number of parameters.

Figure 6.3 depicts the kernel density estimation of the posterior densities for the 10 parameters. The bandwidth in the kernel density estimator is chosen according to Silverman's rule of thumb. The univariate posterior densities are symmetric and resemble normal densities. The conclusions from the posterior inference are very similar to the conclusions in Section 5.4. Note for instance the support for ξ_B, which is mostly negative and agrees with the idea that remembering more words than in the past becomes less likely with increasing age. Effects γ_F and γ_D for *sex* can be considered to be positive given that almost all of the support of the corresponding posterior densities is positive.

The first mover/stayer model, the one with one frailty term for all forward transitions, has DIC = 7690 and $p_D = 309.0$. The second mover/stayer model, with a frailty term for forward transitions through the living states only, has DIC = 7639 and $p_D = 512.5$. In agreement with the marginal likelihood inference in Section 5.4, the mover/stayer models improve upon the fixed-effects model, and the model with the frailty term restricted to the forward transitions through the living states preforms best. For the latter model, the posterior density of σ is depicted in Figure 6.4. The distribution is positively skewed. The posterior mean of σ is 0.999, with 95% credible interval (0.771; 1.268).

It is interesting to investigate characteristics of the individuals in the sample who can be considered to be stayers according to the fitted model. To look into this, order the means of the posterior distributions of b_i, $i = 1, ..., N$, from small to large. The 25 individuals with posterior means in the 2.5% percentile of the ordered means are all observed four or five times, and during the follow-up their average number of words remembered is 6.9, and their average age is 69.3. These are individuals who—despite a higher age—perform very well on the test for cognitive function. This outlier behaviour is nicely captured by the frailty model.

6.5 Inference using the BUGS software

WinBUGS (Lunn et al., 2013) is flexible software for Bayesian inference using MCMC. The acronym BUGS stands for *Bayesian inference Using Gibbs Sampling*. The BUGS-project started in 1989 in the MRC Biostatistics Unit, Cambridge, UK, leading to the WinBUGS software as joint work with the Imperial College School of Medicine at St Mary's, London.

Updated versions of WinBUGS have been released under the name OpenBUGS (Lunn et al., 2013). Models defined in the BUGS language can also be

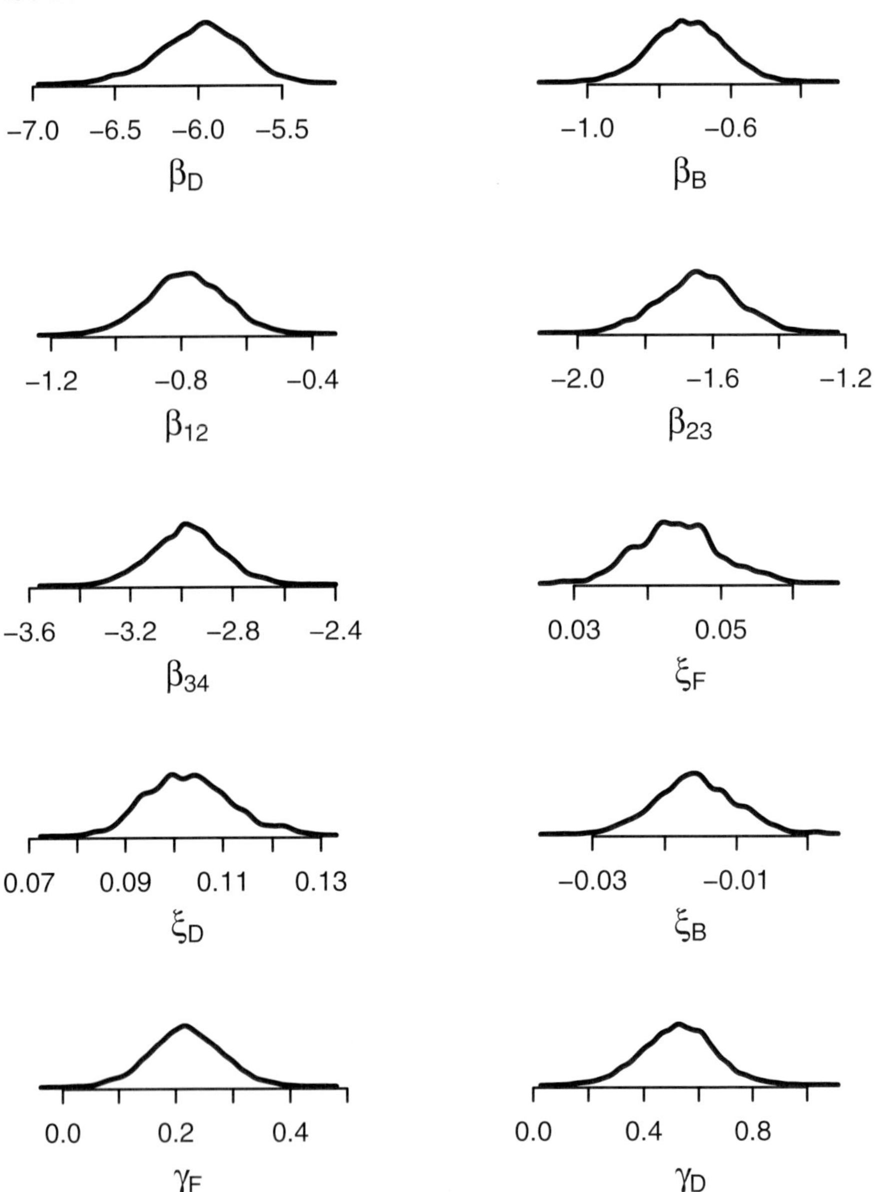

Figure 6.3 *Posterior densities for the 10 parameters in the fixed-effects model for the ELSA data, using kernel density estimation with Silverman's rule of thumb for the bandwidth.*

estimated using JAGS (Plummer, 2003), which stands for *Just Another Gibbs Sampler*. All three computer programs are free.

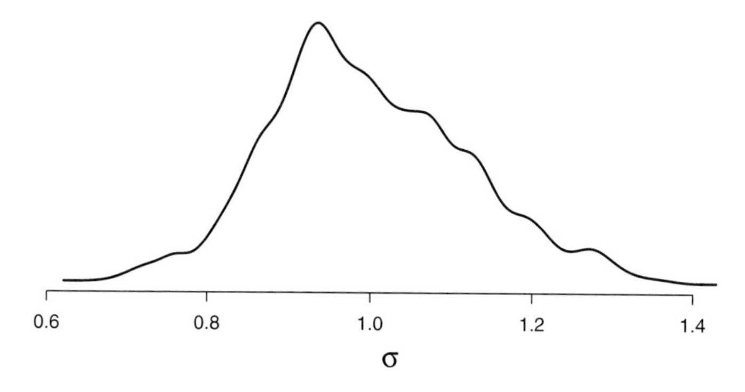

Figure 6.4 *Posterior density for the standard deviation σ in the model for the ELSA data with one log-normal frailty for the forward transitions through the living states. Using kernel density estimation with Silverman's rule of thumb for the bandwidth.*

For many models, the computer programs will produce the same Bayesian inference. For this reason, we will use the term BUGS in a very general way to denote both the language and the implemented sampler, and as an adjective for the model to be estimated. Nevertheless, there are differences between the software. These differences may grow in the future as OpenBUGS and JAGS are still under ongoing development. For the models in the book, we noticed that numerical robustness is important. Because the models for the transition intensities are defined using the exponential function, certain combinations of sampled parameters can lead to large values for the intensities, which can cause numerical overflow. WinBUGS version 1.4.3 turns out to be very robust in this aspect. Since work on WinBUGS has shifted to development of OpenBUGS, the BUGS models in this book are fitted with OpenBUGS version 3.2.3 rev 1012, and with JAGS version 3.4.0.

BUGS allows the user to define models using a coding language that is similar to the language in the R software. There are several advantages of BUGS. The first—and the most important one—is that the user only has to define the model and the prior distributions; BUGS executes the MCMC automatically. A second advantage is that given a BUGS implementation of a fixed-effects model, it is relatively straightforward to extend to modelling by including random effects. A third plus of BUGS is that it is not necessary to define conjugate prior distributions. A conjugate prior distribution for a parameter θ is a distribution that implies a known parametric form for the conditional distribution of θ in the Gibbs sampler. Conjugate prior distributions allow straightforward sampling from conditional distributions which simplified the implementation of Gibbs sampler. On the other hand, using non-conjugate priors often makes it easier for the user to define prior distributions. For variance parameters in hierarchical models, it is even recommended to use non-conjugate uniform distributions when the aim is to define non-informative and weakly informative prior distributions (Gelman, 2006).

In the context of multi-state models, a requirement for using standard versions of BUGS is the availability of a closed-form solution for the matrix with the transition probabilities $\mathbf{P}(t) = \exp(t\mathbf{Q})$. For the standard versions, this closed form has to be implemented explicitly in the BUGS language. For more complex models where the transition probability matrix does not have a closed-form solution, Welton and Ades (2005) show how the WinBUGS Differential Interface (WBDiff) can be used to solve the underlying differential equations numerically within the MCMC simulation in WinBUGS.

The next two sections follow the presentation in Van den Hout and Matthews (2009), which is an extension of the work in Welton and Ades (2005) and Pan et al. (2007). Van den Hout and Matthews (2009) discuss a progressive three-state survival model in BUGS. The same approach is also used in Van den Hout and Tom (2013) for a three-state survival model with recovery.

6.5.1 *Adapted likelihood function*

The likelihood function for the multi-state survival model is presented in Chapter 4. In that specification, transition-specific hazards are included explicitly in the likelihood function to take into account known death times. For example, if the last observation at time t_{J-1} is state y_{j-1}, and death is observed at t_J, then the likelihood contribution of interval $(t_{J-1}, t_J]$ is

$$C(y_J|y_{J-1}, \mathbf{x}) = \sum_{s=1}^{D-1} P(Y_J = s|Y_{J-1} = y_{J-1}, \boldsymbol{\theta}, \mathbf{x})\, q_{sD}(t_{J-1}|\boldsymbol{\theta}, \mathbf{x}), \qquad (6.4)$$

where the notation is the same as in Chapter 4. Expression (6.4) reflects that—due to the interval censoring—the state just before death is unknown and that death from that state is assumed to happen instantaneously with hazard $q_{sD}(t_{J-1}|\boldsymbol{\theta}, \mathbf{x})$.

A disadvantage of using (6.4) for known death times is that the likelihood function is not a probability. Given that the data for observed state are discrete, this can lead to problems when software is used that (quite rightly) assumes that for discrete data the likelihood function is proportional to a probability mass function. BUGS is the motivating example here.

An alternative for (6.4) is

$$C^*(y_J|y_{J-1},\mathbf{x}) = \sum_{s=1}^{D-1} P(\text{state } s \text{ at } t_J - \varepsilon|Y_{J-1} = y_{J-1},\boldsymbol{\theta},\mathbf{x}) \times P(Y_{J-1} = D|\text{state } s \text{ at } t_J - \varepsilon,\boldsymbol{\theta},\mathbf{x}), \quad (6.5)$$

where ε is a user-specified length of time (cf. Kay, 1986). This implies an unknown state just before death at time $t_J - \varepsilon$ followed by death within the time interval $(t_J - \varepsilon, t_J]$. Using (6.5) instead of (6.4) will ensure that the likelihood is a probability mass function. The choice of ε will depend on the application at hand. Sensitivity of the data analysis to the specification of ε can easily be investigated by exploring various values.

6.5.2 *Multinomial distribution*

The following presents distributions for data in a three-state continuous-time survival model. Implementation of a statistical model in BUGS implies defining distributions for the data and defining prior distributions for the parameters. The posterior distribution is not implemented explicitly.

Likelihood expression (6.5) is adopted to enable using BUGS for Bayesian inference. As before, assume that an individual i is observed at times $t_1, t_2, ..., t_J$. The intervals $(t_1,t_2], ..., (t_{J-1},t_J]$ are modelled independently by using the multinomial distribution for the transitions *to* state Y_j given state y_{j-1}, $j = 2, ..., J$. The subscript i is ignored to keep the notation simple.

For the multinomial distribution, state as a response variable is recoded: states 1, 2, and $D = 3$ are coded as (1,0,0), (0,1,0), and (0,0,1), respectively. For this coding, the notation for the response vector is $\overline{\mathbf{Y}} \in \{0,1\}^3$.

We specify a frailty model for the transition intensities with log-normal individual-specific frailties; see Section 5.2. The multivariate frailty for individual i is given by the parameter vector $\mathbf{b}$. As before, small letters are used for random effects as random variables, and for values of random effects.

For $j \in \{2, ..., J\}$, if the individual is in a living state at t_j, then the distribution for time interval $(t_{j-1}, t_j]$ is given by

$$\begin{aligned} \overline{\mathbf{Y}}_j|\, y_{j-1},\boldsymbol{\theta},\mathbf{b} &\sim Multinomial(\boldsymbol{\pi}_j, 1) \\ \mathbf{b}|\, \boldsymbol{\Sigma} &\sim N(\mathbf{0},\boldsymbol{\Sigma}), \end{aligned}$$

where $\boldsymbol{\theta}$ is the vector with the fixed-effects model parameters, and

$$\boldsymbol{\pi}_j = \left(p_{y_{j-1}1}(t_{j-1},t_j), p_{y_{j-1}2}(t_{j-1},t_j), p_{y_{j-1}3}(t_{j-1},t_j)\right), \quad (6.6)$$

where $p_{rs}(t_1,t_2)$ denotes the (r,s) entry of the transition matrix $\mathbf{P}(t_1,t_2)$ defined for interval $(t_1,t_2]$.

In case an individual dies at time t_J, we assume that

$$\overline{\mathbf{Y}}_J|\, y_{J-1},\boldsymbol{\theta},\mathbf{b} \;\sim\; Multinomial(\boldsymbol{\pi}_J,1),$$

where

$$\boldsymbol{\pi}_J = \left(\frac{p_{y_{J-1}1}(t_{J-1},t_J)(1-\pi_d)}{1-p_{y_{J-1}3}(t_{J-1},t_J)}, \frac{p_{y_{J-1}2}(t_{J-1},t_J)(1-\pi_d)}{1-p_{y_{J-1}3}(t_{J-1},t_J)}, \pi_d\right) \tag{6.7}$$

with

$$\begin{aligned} \pi_d &= p_{y_{J-1}1}(t_{J-1},t_J-\varepsilon)\times p_{13}(t_J-\varepsilon,t_J) \\ &+ p_{y_{J-1}2}(t_{J-1},t_J-\varepsilon)\times p_{23}(t_J-\varepsilon,t_J). \end{aligned} \tag{6.8}$$

This means that in case of death at t_J, we assume an unknown state just before death and then a transition into the death state within a small time interval ε, as implied by the likelihood contribution (6.4). The probability of this event is denoted π_d. Because of the ε-approximation of the exact death time in (6.8), it follows that $1-\pi_d \neq p_{y_{J-1}1}(t_{J-1},t_J)+p_{y_{J-1}2}(t_{J-1},t_J)$, hence the adjustment in (6.7) to ensure a proper distribution.

6.5.3 *Right censoring*

In the presence of right censoring, the multinomial distribution in the previous section is extended. States 1, 2, 3, and the censored state are coded by the vectors (1,0,0,0), (0,1,0,0), (0,0,1,0), and (0,0,0,1), respectively. The vectors are denoted by $\overline{\mathbf{Y}}^* \in \{0,1\}^4$. If the state at t_j is neither dead nor right-censored, we assume that

$$\begin{aligned} \overline{\mathbf{Y}}_j^*|\, y_{j-1},\theta,\mathbf{b} &\sim Multinomial(\boldsymbol{\pi}_j,1) \\ \mathbf{b}|\,\boldsymbol{\Sigma} &\sim N(\mathbf{0},\boldsymbol{\Sigma}), \end{aligned}$$

where

$$\boldsymbol{\pi}_j = (p_{y_{j-1}1}(t_{j-1},t_j), p_{y_{j-1}2}(t_{j-1},t_j), p_{y_{j-1}3}(t_{j-1},t_j), 0)$$

for $j=2,...,J$. If the state at time t_J is the dead state or a censored state we assume that

$$\overline{\mathbf{Y}}_J^*|\, y_{J-1},\boldsymbol{\theta},\mathbf{b} \;\sim\; Multinomial(\boldsymbol{\pi}_J^*,1),$$

where, in case of death, we define $\boldsymbol{\pi}_J^* = (\boldsymbol{\pi}_J, 0)$ with $\boldsymbol{\pi}_J$ defined by (6.7). In case of a censored state, we define

$$\boldsymbol{\pi}_J^* = \big(0, 0, p_{y_{J-1}3}(t_{J-1}, t_J), 1 - p_{y_{J-1}3}(t_{J-1}, t_J)\big).$$

This specifies the contribution to the likelihood as $1 - p_{y_{J-1}3}(t_{J-1}, t_J)$, which is the probability of not being dead at t_J given an observed state at t_{J-1}. This is in line with the definition of the likelihood contribution in (6.4).

6.5.4 *Example: frailty in the Parkinson's disease study*

The Parkinson's disease data are introduced in detail in Section 3.7.1. Data are available for $N = 233$ individuals with Parkinson's disease. The states are defined as 1 = dementia-free, 2 = dementia, and 3 = dead. For the progressive model, Section 3.7.3 shows that—according to AIC—a three-state model with Weibull baseline hazards fits better than a model with Gompertz baselines. We will define BUGS models for the Parkinson's disease data and reassess the choice of baseline hazards. In addition, a frailty model will be defined to investigate mitigation of the Markov assumption.

The likelihood function will be derived from the distributional assumptions in Section 6.5.2. This implies that the DIC values in the current section cannot be compared to the values in Section 6.3.

The piecewise-constant approximation of parametric time dependency is used, and the time intervals in the data are used to define the grid; see Section 4.4. This means that given two observed states at consecutive times t_1 and t_2, generator matrix $\mathbf{Q}(t_1)$ is defined at t_1 and kept constant within the interval $(t_1, t_2]$. For the progressive three-state survival model, the 3×3 matrix with the transition probabilities $\mathbf{P}(t_1, t_2) = \exp\big((t_2 - t_1)\mathbf{Q}(t_1)\big)$ can be expressed in closed form; see Section 3.2.1 and Appendix A. Because of this, the computation of $\mathbf{P}(t_1, t_2)$ can be explicitly implemented in BUGS language.

First we investigate the fixed-effects Gompertz model defined by

$$\begin{aligned}
q_{12}(t) &= \exp(\beta_{12.0} + \xi_{12}\, t + \beta_{12.s}\, sex + \beta_{12.d}\, duration) \\
q_{13}(t) &= \exp(\beta_{13.0}) \\
q_{23}(t) &= \exp(\beta_{23.0} + \xi_{23}\, t + \beta_{23.s}\, sex + \beta_{23.d}\, duration),
\end{aligned}$$

and the Weibull variant

$$\begin{aligned}
q_{12}(t) &= \tau_{12} t^{\tau_{12}} \exp(\beta_{12.0} + \beta_{12.s}\, sex + \beta_{12.d}\, duration) \\
q_{13}(t) &= \exp(\beta_{13.0}) \\
q_{23}(t) &= \tau_{23} t^{\tau_{23}} \exp(\beta_{23.0} + \beta_{23.s}\, sex + \beta_{23.d}\, duration).
\end{aligned}$$

Table 6.3 *Weibull three-state progressive models for the Parkinson's disease data. Posterior means and 95% credible intervals. The frailty model is defined by including a random intercept in the log-linear regression for transition* $2 \to 3$.

	Fixed-effects model		Frailty model	
$\beta_{12.0}$	−13.540	(−14.950; −10.640)	−15.800	(−19.410; −11.590)
$\beta_{13.0}$	−4.574	(−8.450; −3.487)	−4.833	(−13.320; −3.477)
$\beta_{23.0}$	−9.938	(−12.530; −5.639)	−11.120	(−17.880; −5.164)
τ_{12}	3.763	(3.035; 4.154)	4.331	(3.271; 5.242)
τ_{23}	3.046	(2.011; 3.678)	3.351	(1.916; 5.025)
$\beta_{12.s}$	−0.519	(−0.905; −0.140)	−0.554	(−0.940; −0.174)
$\beta_{23.s}$	−0.372	(−0.685; −0.054)	−0.406	(−0.766; −0.061)
$\beta_{12.d}$	0.064	(0.024; 0.103)	0.069	(0.029; 0.110)
$\beta_{23.d}$	−0.005	(−0.034; 0.022)	−0.006	(−0.037; 0.024)
σ			0.320	(0.050; 0.641)

As in Section 3.7.3, the time scale t in these models is age minus 35, which results in 1 being the smallest value of t in the data. The restrictions on the parameters for transition $1 \to 3$ are the same as in the final Weibull model in Section 3.7.3.

For the BUGS models, specifications of prior distributions for the parameter vectors are required. We used the following specifications for the Gompertz model:

$$\begin{aligned} \beta_{12.0}, \beta_{13.0}, \beta_{23.0} &\sim U(-20, 10) \\ \beta_{12.s}, \beta_{23.s}, \beta_{12.d}, \beta_{23.d} &\sim U(-20, 10) \\ \xi_{12}, \xi_{23} &\sim U(-10, 10), \end{aligned}$$

and for the Weibull we use $\tau_{12}, \tau_{23} \sim U(0, 10)$, *ceteris paribus*. Specifying univariate prior densities for each of the parameters implies *a priori* independence. This does not imply *a posteriori* independence.

Both fixed-effects models have the same number of parameters. The DIC for the Gompertz model is 2251, with $p_D = 8.6$. For the Weibull we have DIC = 2247, with $p_D = 7.8$. As in Section 3.7.3, the Weibull performs better, but the difference is rather small.

See Table 6.3 for the posterior inference of the parameters in the Weibull model. The Bayesian inference in this table can be compared with the maximum likelihood inference in Table 3.5. There is some discrepancy between the posterior means for $\beta_{12.0}$ and τ_{12}, and their maximum likelihood counterparts. The posterior distribution of these two parameters for the transition

$1 \to 2$ are strongly (negatively) correlated. But overall the differences with the maximum likelihood inference do not lead to a difference assessment of the multi-state process. Note for instance that the posterior means for the effect of the covariates are very close to maximum likelihood counterparts.

Because the running of the Gibbs sampler is automated in BUGS, it is relatively easy to change a model or to extend it. Exploring another choice for the baseline hazard only involves changing the coding of the hazard regression models and the definition of the prior distributions. In Chapter 4, the log-logistic hazard model is mentioned as an alternative to the Gompertz and the Weibull. For the data at hand, BUGS code was changed to fit the model

$$
\begin{aligned}
q_{12}(t) &= \frac{\lambda_{12}\rho_{12}(\lambda_{12}t)^{\rho_{12}-1}}{1+(\lambda_{12}t)^{\rho_{12}}} \exp(\beta_{12.s}\, sex + \beta_{12.d}\, duration) \\
q_{13}(t) &= \lambda_{13} \\
q_{23}(t) &= \frac{\lambda_{23}\rho_{23}(\lambda_{23}t)^{\rho_{23}-1}}{1+(\lambda_{23}t)^{\rho_{23}}} \exp(\beta_{23.s}\, sex + \beta_{23.d}\, duration),
\end{aligned}
$$

with prior distributions defined by

$$
\begin{aligned}
\lambda_{12}, \lambda_{13} &\sim U(0,2) \\
\lambda_{23} &\sim U(0,0.05) \\
\beta_{12.s}, \beta_{23.s}, \beta_{12.d}, \beta_{23.d} &\sim U(-20,10) \\
\rho_{12}, \rho_{23} &\sim U(0,20),
\end{aligned}
\tag{6.9}
$$

The choice for the prior distributions for the λ-parameters are partly motivated by numerical reasons: large values of the λ-parameters are not to be expected, but—as was noted in some preliminary sampling—have a disturbing effect on the behaviour of the MCMC chains. Using $U(0,2)$ as the prior distribution for λ_{23} resulted in an MCMC for this parameter with short episodes of sampled values close to 2. This sampling of large values introduces spikes in the MCMC chains leading to a very non-normal posterior distribution of λ_{23}, which affects the computation of the DIC to such an extent that the number of effective parameters is estimated at a negative value. Using the prior distributions in (6.9), the posterior means (and 95% credible intervals) for the λ-parameters are

$$
\begin{aligned}
\widehat{\lambda}_{12} &= 0.026 \;\; (0.024; 0.028) \\
\widehat{\lambda}_{13} &= 0.021 \;\; (0.008; 0.036) \qquad \widehat{\lambda}_{23} = 0.028 \;\; (0.026; 0.031),
\end{aligned}
$$

which do not conflict with the support enforced by choice of the prior distributions. The model has DIC = 2261 with $p_D = 8.8$. Hence the log-logistic model does not perform better than the Weibull model.

Changing a fixed-effects BUGS model into a frailty model implies specifying the distribution for the frailties and including the frailties in the regression models for the hazards. The automated Gibbs sampler will sample each frailty from its distribution conditional on the other parameters. For the Parkinson's disease data with $N = 233$ patients, introducing an individual-specific univariate frailty implies sampling the parameters for the frailty distribution *and* the 233 individual-specific frailties. Understandably, the execution time of the BUGS software will increase considerably when switching from fixed-effects models to models with random effects. This will be illustrated below.

As an example, we extend the Weibull model by adding individual-specific frailties $b_i \sim N(0, \sigma)$ to the regression model for transition $2 \rightarrow 3$. We define

$$\begin{aligned} q_{12}(t|i) &= \tau_{12} t^{\tau_{12}} \exp(\beta_{12.0} + \beta_{12.s}\, sex + \beta_{12.d}\, duration) \\ q_{13}(t|i) &= \exp(\beta_{13.0}) \\ q_{23}(t|i) &= \tau_{23} t^{\tau_{23}} \exp(\beta_{23.0} + b_i + \beta_{23.s}\, sex + \beta_{23.d}\, duration). \quad (6.10) \end{aligned}$$

As stated at the start of Chapter 5, a possible motivation for introducing a frailty term is to counterbalance violations of the conditional Markov assumption. In the current example, if duration of stay in the dementia state is correlated with the hazard of a transition to the dead state, then this would constitute such a violation. By including b_i, model (6.10) allows for heterogeneity beyond observed age and covariate values. If there is a strong effect of duration of stay in the dementia state which is not informed by age or covariates, then the individual-specific frailty term may be able to distinguish patients who differ with respect to this duration.

For the fixed-effects parameters in the Weibull frailty model (6.10), the prior distributions are the same as the ones for the fixed-effects model. The prior distribution for the standard deviation of the frailty distribution is $\sigma \sim U(0, 5)$.

When adding a frailty term to a fixed-effects model, it is to be expected that the execution time of the MCMC sampling increases. First, each sample iteration takes more time as there are more parameter values to sample. However, for BUGS, this increase in time is rather minimal. For the same number of iterations, the frailty model only took 7% longer compared to the fixed-effects model. Second, due to the increased model complexity it may take more iterations to obtain MCMC chains that are stable. This is indeed distinctive for the Weibull models: for the fixed-effects model two chains of each 10,000 iterations suffice, whereas the frailty model needs about 30,000 iterations for each of the two chains.

The expected deviance $\overline{D}$ for the Weibul frailty model is 2224.0 and smaller than the expected deviance $\overline{D} = 2240.0$ for the fixed-effects Weibull model. This is as expected since the frailty model is more complex than the fixed-effects model. However, taking into account the number of effective parameters, the DICs for the models are about the same. For the frailty model, DIC = 2248 ($p_D = 22.5$), and for fixed-effects model DIC = 2247 ($p_D = 7.7$). Table 6.3 presents the posterior inference of the parameters in the frailty model. Although there are differences in the posterior means when comparing the results of the fixed-effects model, the overall inference for the fixed-effects parameters is similar when the two models are compared.

Seen as a check of the conditional Markov assumption, this comparison of DICs suggests that the assumption is not violated. Further insight can be gained by looking at the posterior distributions of the $N = 233$ frailties. Figure 6.5 illustrates the variation in the 233 posterior means (top panel), and the posterior distribution for each of the 233 frailty terms (boxplots in bottom panel). The four patients with a mean of the posterior frailty distribution smaller than 0.5 are the patients with indexes 4, 32, 75, and 102; see the boxplots in the bottom panel. These patients have been observed in the dementia state more often than other patients in the data. The posterior means for the frailty term for these patients are negative, which implies a correction for the risk of moving as determined by the fixed effects. In other words, according to the fixed effects they should have moved but this is counterbalanced by negatively valued frailties.

Even though the frailty term is able to pick up heterogeneity not captured by the covariates (that is, prolonged duration of stay in the dementia state), adding the frailty term to the fixed-effects model does not lead to a better fit according to the DIC.

An extended model was fitted by including the frailty term b_i in the log-linear regression for transition $1 \to 2$ as well as in the regression for transition $2 \to 3$. This model can be interpreted as a mover/stayer model in the sense that individuals with higher value of b_i tend to move quicker through the trajectory $1 \to 2 \to 3$. With DIC = 2247, this extended model did not perform better than the previous model.

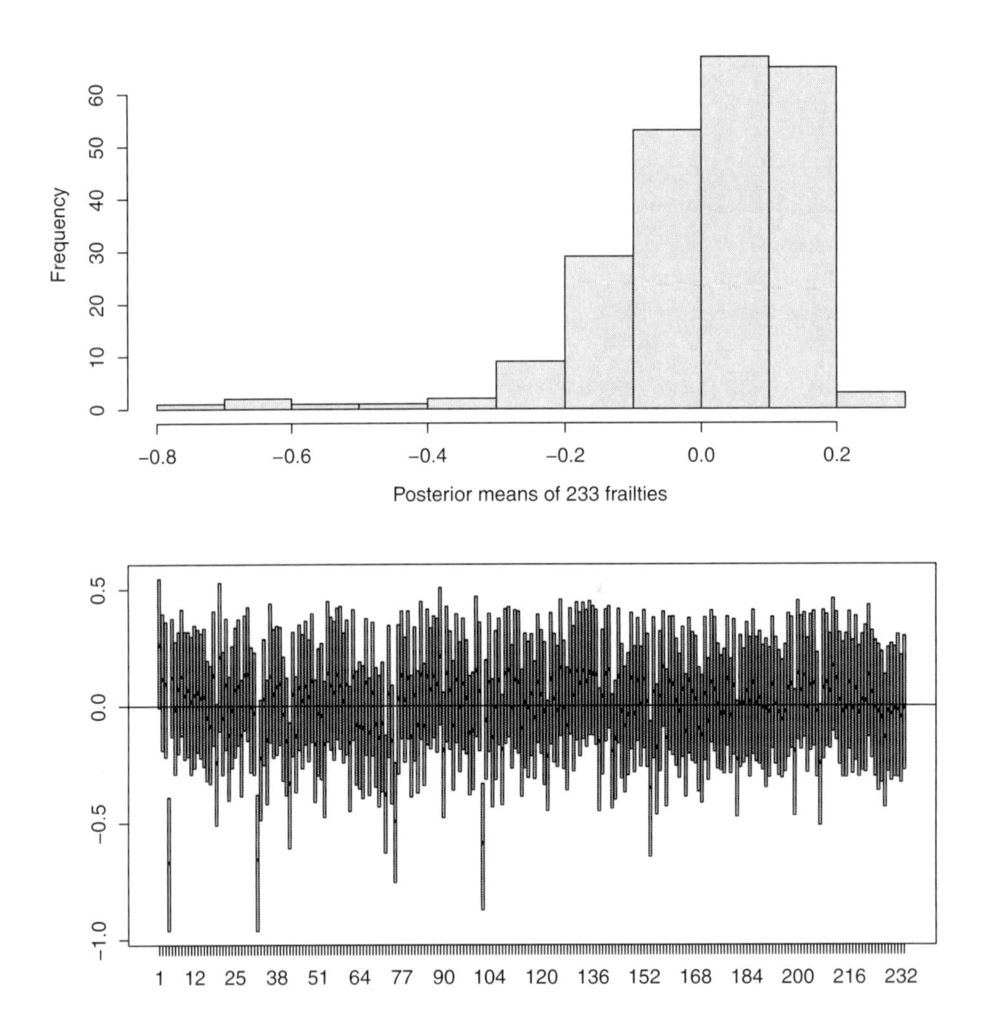

Figure 6.5 *Posterior inference for the 233 individual-specific frailties in the Weibull hazard model for the transition from dementia to dead. Bottom panel, boxplots for each of the 233 frailties with first, second, and third quartiles.*

Chapter 7

Residual State-Specific Life Expectancy

7.1 Introduction

Residual life expectancy is the expected number of years of life remaining at a given age. Given a multi-state survival model, total residual life expectancy can be distinguished from residual life expectancy in a specific state. State-specific life expectancies are of importance in ageing research, where not only longevity is of interest but also whether individuals spent their old age in good health.

Total and state-specific life expectancies can be derived from an estimated multi-state survival model as functions of model parameters. Because life expectancies are computed using transition probabilities, the basic approach is the same for discrete-time models and continuous-time models; see, for example, Lièvre et al. (2003) for the former and Izmirlian et al. (2000) for the latter. State-specific life expectancy is an example of a *state occupation time*. The latter is a more general name for duration of stay in a state; see, for example, Willekens (2014, Section 2.4). The life-expectancy terminology is typically used when the time scale is age in years (or can be linked directly to age in years), and when the specified process includes an absorbing dead state.

Life expectancies can also be estimated from life tables. A life table (or an actuarial table) shows, for a range of ages, the probability that a person of a certain age will die within a year. Typically, life tables are derived from cross-sectional data. State-specific life expectancies can be estimated from a life table with the method introduced by Sullivan (1971); see also Jagger et al. (2007) and Imai and Soneji (2007). Because the current book is on the analysis of longitudinal data, the Sullivan method will not be discussed in what follows. A comparison between the two approaches is discussed in Newman (1988).

After introducing the basic definitions in Section 7.2, this chapter discusses computation of life expectancies by integration in Section 7.3, and computation by micro-simulation in Section 7.5. The methods are illustrated and compared in Sections 7.4 and 7.6.

7.2 Definitions and data considerations

This section will derive the expressions for total and state-specific life expectancies. Once parameters of a multi-state survival model are estimated, these expressions can be used for inference on years of life remaining at a given age. At the end of the section, some data considerations are discussed.

An example with a three-state survival model may help to understand state-specific life expectancy as mean survival in a specified state. With state $D = 3$ the dead state, consider the progressive time-homogeneous model defined by transition intensities

$$\begin{aligned} q_{12}(t) &= \lambda_{12} \\ q_{13}(t) = q_{23}(t) &= \lambda_D \qquad \text{for} \quad \lambda_{12}, \lambda_D > 0, \end{aligned}$$

where $q_{rs}(t)$ is the intensity (hazard) of moving from state r to state s at time t. This is an exponential model. Let T_1 be the event time of leaving state 1. From the properties of the exponential distribution, it follows that $T_1 \sim Exp(\lambda_{12} + \lambda_D)$ and that mean survival in state 1 is $\mathbb{E}(T_1) = 1/(\lambda_{12} + \lambda_D)$. The hazard of dying is the same for state 1 and state 2, so for T defined as the time of death, overall mean survival is given by $\mathbb{E}(T) = 1/\lambda_D$. With this, mean survival in state 2 is given by $\mathbb{E}(T) - \mathbb{E}(T_1) = \lambda_{12}/(\lambda_D \lambda_{12} + \lambda_D^2)$.

For an interpretation in terms of life expectancies, let the time scale of the model be in years. If an individual is in state 1 at age t, then total residual life expectancy is given by $\mathbb{E}(T)$, expected number of years spent in state 1 is given by $\mathbb{E}(T_1)$, and expected number of years in state 2 is the difference.

Since there is no age effect in the model for the transition intensities, there is no age effect for the residual life expectancies. In practice, this will often not be realistic. Also, the assumption that the hazard of dying is the same for state 1 and state 2 will often be violated in practice. The remainder of the chapter is about the definition and computation of life expectancies given extended, time-dependent models.

To introduce the concept of residual life expectancy formally, we start with the two-state model where state 1 is the living state and state 2 is the dead state. This is the standard survival model from Chapter 2. Survival time

T is the time spent in state 1. It follows that

$$\begin{aligned}
\mathbb{E}(T) &= \int_0^\infty t f(t)dt = \int_0^\infty \left[\int_0^\infty I(t>u)du\right] f(t)dt \\
&= \int_0^\infty \left[\int_0^\infty I(t>u)f(t)dt\right] du \\
&= \int_0^\infty \mathbb{E}\left[I(T>u)\right] du = \int_0^\infty P\big(I(T>u)=1\big)du \\
&= \int_0^\infty S(u)du,
\end{aligned}$$

where $I()$ is the indicator function; see also Aalen et al. (2008, Section 3.2.3). The choice of the probability density function f is of course essential and different choices of f can induce large differences in mean survival time.

Assume that the time scale is years and that the two-state process is dependent on covariate values. Denote the covariate values at time t by $\mathbf{x}(t)$ and consider $\mathcal{X}_{t_1} = \{\mathbf{x}(t) \mid t \geq t_1\}$ to be known for given age t_1. Total residual life expectancy at age t_1 conditional on $\mathcal{X}_{t_1}$ is now given by the expected remaining years spent in state 1; that is, by

$$\mathbb{E}(T|t_1,\mathcal{X}_{t_1}) = \int_0^\infty S(t_1+u|t_1,\mathcal{X}_{t_1})du = \int_0^\infty P(Y_{t_1+u}=1|Y_{t_1}=1,\mathcal{X}_{t_1})du,$$

where $Y_t \in \{1,2\}$ is the random variable denoting the state at time t.

From the above, it is clear that mean survival time is only finite when $\int_0^\infty S(u)du$ is finite. An example where this is not the case is $T \sim$ *Gompertz*(λ,ξ) with $\xi < 0$. As already stated in Chapter 2, if $\xi < 0$, then for t very large, the survivor function goes to $\exp(\lambda\xi^{-1}) > 0$. Hence, when a Gompertz hazard is fitted with $\widehat{\xi} > 0$, total residual life expectancy cannot be inferred from the definition of $\mathbb{E}(T|t_1,\mathcal{X}_{t_1})$.

In a multi-state survival process, there is more than one living state. We use the notation introduced in Chapter 4 for the continuous-time process $\{Y_t|t \geq 0\}$ with state space $S = \{1,...,D\}$, where D is the dead state. The definition of residual life expectancy in a specific living state is a direct extension of the definition for the two-state survival model.

Denote time in state s by T_s. It follows that

$$\begin{aligned}
\mathbb{E}\left[T_s \middle| Y_0=r\right] &= \mathbb{E}\left[\int_0^\infty I(Y_t=s)dt \middle| Y_0=r\right] \\
&= \int_0^\infty \mathbb{E}\left[I(Y_t=s)\middle| Y_0=r\right] dt = \int_0^\infty P(Y_t=s|Y_0=r)dt;
\end{aligned}$$

see Kulkarni (2011, Theorem 4.4).

If the multi-state process at time t is dependent on covariate vector $\mathbf{x}(t)$, then assume that $\mathcal{X}_{t_1} = \{\mathbf{x}(t) | \ t \geq t_1\}$ is known. Given age as the time scale, residual life expectancy in living state s given state r at age t_1, for $r, s \in \{1, 2, ..., D-1\}$, is now defined by

$$e_{rs}(t_1) = e_{rs}(t_1 | \mathcal{X}_{t_1}) = \int_0^{\infty} P(Y_{t_1+u} = s | \ Y_{t_1} = r, \mathcal{X}_{t_1}) du, \tag{7.1}$$

where $P(Y_{t_1+u} = s | \ Y_{t_1} = r, \mathcal{X}_{t_1})$ is the transition probability of being in state s at age $t_1 + u$, conditional on starting state r at age t_1 and covariate process $\mathcal{X}_{t_1}$.

If $r \neq s$ in (7.1), then $e_{rs}(t_1)$ does not include any specification of time spent in r or in states other than state s. Say there are four living states. If $e_{14}(t_1)$ is 2 years, then this means only that starting in state 1 at specified age t_1, the expected time spent in state 4 is 2 years. How the rest of the time spent in the other living states is divided over these states can only be derived from additional computation of $e_{11}(t_1)$, $e_{12}(t_1)$, and $e_{13}(t_1)$.

A useful definition is the marginal residual life expectancy given by

$$e_{\bullet s}(t_1) = \sum_{r \neq D} e_{rs}(t_1) P(Y_{t_1} = r | \mathcal{X}_{t_1}), \tag{7.2}$$

where the summation is over all living states $r \in \{1, 2, ..., D-1\}$. Marginal life expectancy in state s is the expected number of years spent in state s irrespective of the initial state at age t_1. To derive this quantity we need the distribution of the living states at age t_1; that is, we need $P(Y_{t_1} = r | \mathcal{X}_{t_1})$ for all $r \in \{1, 2, ..., D-1\}$. This distribution can be estimated by fitting a multinomial logistic regression model where the data stem from the observations of the living states in the multi-state process. This estimation can be undertaken independently from the fitting of the multi-state model.

Total residual life expectancy at age t_1 is defined as

$$e(t_1) = \sum_{s \neq D} e_{\bullet s}(t_1). \tag{7.3}$$

Quantities (7.1), (7.2), and (7.3) can be derived from the fitted multi-state model and an additionally fitted multinomial logistic regression model. The specification of $\mathcal{X}_{t_1}$ is of course important, as is the modelling of the time dependency of the multi-state process.

To approximate the integral (7.1) up to infinity, a maximum age t_{max} has to be specified such that we may safely assume that the integrand $P(Y_{t_1+u} = s | \ Y_{t_1} = r, \mathcal{X}_{t_1})$ is 0 for $s \neq D$ when $t_1 + u > t_{max}$.

Given a fitted multi-state model, estimation of (7.1), (7.2), and (7.3) is relatively straightforward if $\mathcal{X}_{t_1}$ is known. This is the case when covariates are either time-independent or when the time dependency is external. The following sections will present and illustrate two methods for the estimation of residual life expectancies: integration and micro-simulation. The latter is especially useful in the presence of internal time-dependent covariates.

The methods in this chapter are about estimating residual life expectancy from a fitted multi-state survival model. If gender is a covariate in the model, then the estimation can differentiate in life expectancies for men and women. However, this is only a numerical illustration of the gender effect. The methods do not provide a direct estimate of such an effect. For a competing risks process, Andersen (2013) proposes a regression model for direct effects of covariates on the expected number of years lost due to one of the competing causes. This is closely related to the topic in this chapter since the *expected number of years lost before time* τ is defined as $L(0,\tau) = \tau - \int_0^\tau S(t)dt$. The integral $\int_0^\tau S(t)dt$ is called the restricted mean lifetime, and can be of interest in its own right; see, for example, Zhang and Schaubel (2011).

Once parameters of a multi-state survival model are estimated, the definitions above can be used for inference on years of life remaining at a given age. However, the estimation of the residual life expectancies is only as good as the estimated multi-state model.

First, if there is bias in the estimated model, then this bias is propagated in the estimation of the life expectancies. Heuristically speaking, minor bias in the estimated effect of an explanatory variable on a transition hazard may be innocuous as long as one investigates only whether there is an association between the variable and the hazard. But when life expectancies are calculated for different values of the explanatory variable, minor bias in the estimation of the effect will be inflated due to the propagation in the calculation, and inference on life expectancies may be biased more severely.

Second, even if there is no bias in the estimated model, one has to take into account the age range in the data to which the model is fitted. In most cases, the definition of the life expectancies implies an extrapolation of the model beyond the study time. As an example, say a longitudinal study is set up where all individuals are 75 years old at baseline. If the follow-up time is 15 years, then estimated life expectancies are based on extrapolation of the model beyond the age scale in the data. In addition, if the study started in 2000, and the results are used to compute life expectancies for individuals who are 75 years of age in 2015, then there is also an extrapolation across birth cohorts.

Note also that in this example, the estimation of $P(Y_{t_1} = r|\mathcal{X}_{t_1})$, which is needed for (7.2), is a problem for t_1 defined as 75 years old in 2015. Data for the initial-state distribution in 2015 may not be available.

7.3 Computation: integration

Given a fitted multi-state survival model and specified $\mathcal{X}_{t_1}$, the integrand in (7.1) can be computed for any $u > 0$. Numerical approximation of the integral can thus be used to compute residual life expectancies.

In the computation of this integrand, a piecewise-constant approximation can be used to account for the time dependency of the transition intensities. In that case, it is convenient computationally to use the same grid for the piecewise-constant approximation *and* for the numerical approximation of the integral.

As an aside, the piecewise-constant approximation in the computation of the life expectancies is a numerical approximation of a function of model parameters, whereas using a piecewise-constant approximation in the fitting of the model (see Section 3.4) is based on an assumption with respect to the data and the behaviour of the process. If this assumption does not hold, the estimation of life expectancies will be biased irrespective of the chosen grid for computing (7.1). This illustrates again that the estimation of the residual life expectancies is only as good as the estimated multi-state model. If the piecewise-constant approximation in the fitting of the model leads to bias inference with respect to the model parameters, then this bias is propagated in the estimation of the life expectancies.

The numerical approximation of the integral will provide a point estimate of residual life expectancies. Asymptotic properties of the maximum likelihood estimator of the parameters for the multi-state model can be used to obtain standard errors or confidence intervals; see Section 4.5.

A suite of R functions was written by the author to undertake the above estimation of residual life expectancies for a multi-state model fitted using msm (Jackson, 2011). This model should have Gompertz baseline hazards fitted in msm using the piecewise-constant approximation explained in Section 3.4. The suite of functions is called ELECT. Within ELECT a multinomial logistic regression model is fitted to derive the marginal life expectancies in (7.2).

The methods in ELECT are based on the above theory: the integral in (7.1) is approximated numerically, and a multinomial logistic regression model is fitted to compute (7.2) and (7.3). This will be illustrated in Section 7.6.

An alternative for obtaining standard errors or confidence intervals is using a non-parametric bootstrap. In this case, a number of random samples of individuals with replacement is taken from the data. Because of the sampling with replacement, data of an individual can be included more than once in a bootstrap sample. For each of the samples, the multi-state model is estimated and residual life expectancies are computed using the estimated model. The variation in the estimation of the life expectancies can then be derived. This method takes into account model uncertainty, but is computationally intensive.

7.4 Example: a three-state survival process

Given model parameters for a progressive three-state survival process, this section illustrates the computation of life expectancies using numerical integration. Once the parameters of multi-state survival model are estimated, the data are not used in the computation of life expectancies. To stress this, values of model parameters will be specified in this section without any data. To be able to compute total life expectancy at a given age irrespective of the state at that age, parameters of a logistic regression model are also specified.

For a three-state progressive Gompertz model $q_{rs} = \exp(\beta_{rs} + \xi_{rs}t)$, with t age in years, the parameter values are fixed as

$$\begin{aligned} \beta_{12} &= -4.2 & \xi_{12} &= 0.020 \\ \beta_{13} &= -4.0 & \xi_{13} &= 0.005 \\ \beta_{23} &= -3.0 & \xi_{23} &= 0.015. \end{aligned}$$

For the logistic model

$$\text{logit}(P(Y_{t_1} = 1|t_1) = \gamma_0 + \gamma_1 t_1,$$

the parameters are fixed as $(\gamma_0, \gamma_1) = (6.0, -0.07)$.

To compute life expectancies given specified current age t_1, a grid is needed to approximate the integral in (7.1). The grid will be defined equidistantly by $t_j = t_{j-1} + h$, for $j = 2, ..., J$ and $h > 0$. We will start with a 1-year grid to illustrate the method, and will switch to a 1-month grid for the final computation. To fix ideas, say $t_1 = 65$ and maximum age is $t_{max} = 115$. The 1-year grid is defined by taking $h = 1$, which implies in the current setting that $J = 51$.

For each time interval (t_{j-1}, t_j), $j = 2, ..., J$, a 3×3 generator matrix $\mathbf{Q}(t_{j-1})$ can be computed using the model parameters specified above. With

$\mathbf{Q}(t_{j-1})$ constant throughout the interval $(t_{j-1}, t_j]$, transition probability matrix $\mathbf{P}(t_{j-1}, t_j)$ can be computed as well. As a next step, transition probability matrices can be computed for all intervals $(t_1, t_j]$, $j = 2, ..., J$. Consider, for example, the interval $(t_1, t_3] = (65, 67]$. Using the 1-year grid, first compute $\mathbf{P}(65, 66)$ and $\mathbf{P}(66, 67)$. Next, the transition probability matrix for $(t_1, t_3]$ is approximated piecewise-constantly by the matrix multiplication $\mathbf{P}(65, 66) \times \mathbf{P}(66, 67)$.

In general, for any time interval $(t_1, t_j]$, the transition probability matrix $\mathbf{P}(t_1, t_j)$ is approximated by

$$\mathbf{P}(t_1, t_2) \times \mathbf{P}(t_2, t_3) \times ... \times \mathbf{P}(t_{j-1}, t_j).$$

This computation is illustrated in the top two panels of Figure 7.1. For the choice $t_1 = 65$ and the 1-year grid, the grey lines connect the computed entries of $\mathbf{P}(t_1, t_j)$, for each of the values $t_j = 66, 67, ..., 115$. The top panel of Figure 7.1 depicts the transition probabilities $p_{11}(t_1, t_j)$, $p_{12}(t_1, t_j)$, and $p_{13}(t_1, t_j)$ for changing j; the middle panel depicts $p_{22}(t_1, t_j)$ and $p_{23}(t_1, t_j)$. In line with many three-state models for ageing processes in the older population, the probability to move to state 2 increases initially with age, but then decreases as the hazard of death takes over, and moving from state 1 to the dead state becomes more likely than moving from state 1 to state 2.

The 1-year grid is quite coarse. Additional computation with a 1-month grid, that is, with $h = 1/12$, is also shown in Figure 7.1. For the latter grid, the bottom panel in Figure 7.1 depicts the probability to be in state 1 for any of the ages defined in the 1-month grid. This bottom graph is derived from the fixed values for (γ_0, γ_1). The three panels taken together represent all the information that is needed to compute the residual life expectancies conditional on $t_1 = 65$.

The state-specific life expectancies $e_{rs}(t)$ in (7.1) are defined for $(r, s) \in \{(1,1), (1,2), (2,2)\}$ by the areas under the corresponding curves in Figure 7.1.

Using the grid defined by $h = 1/12$ and Simpson's rule for approximating the integral in (7.1), we obtain for $t_1 = 65$ the following life expectancies:

$$\widehat{e}_{11}(t_1) = 10.63, \qquad \widehat{e}_{12}(t_1) = 4.25, \quad \text{and} \quad \widehat{e}_{22}(t_1) = 6.87.$$

Even though there is no parameter uncertainty in this example, we still use the hat-notation to stress that these figures will always be estimates in practice.

For the marginal and the total life expectancies, we need to use the logistic regression model. With the fixed values for (γ_0, γ_1), we derive that $P(Y_{t_1} = 1 | t_1 = 65)$ is equal to 0.81; see also Figure 7.1.

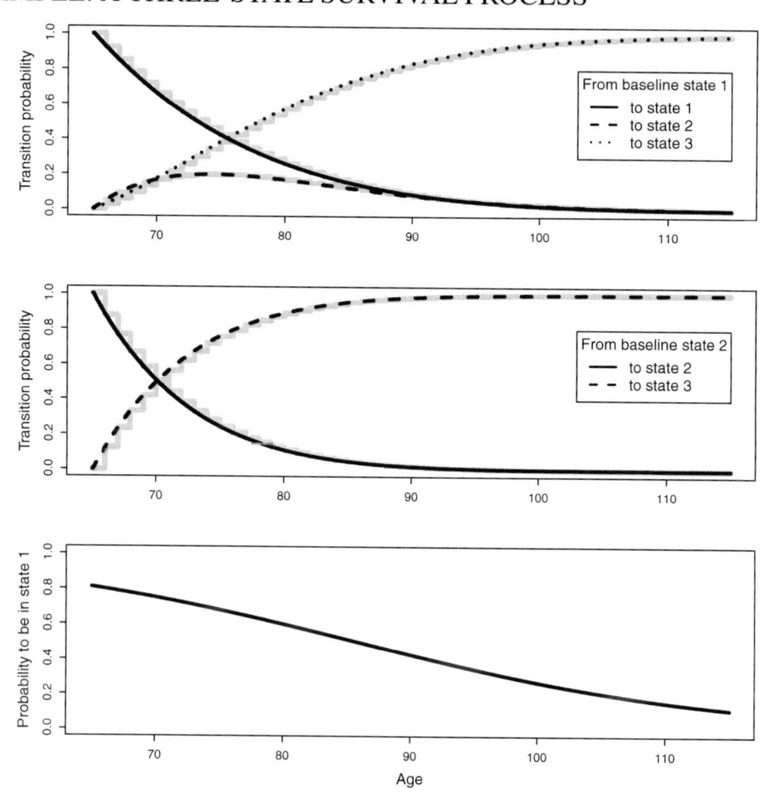

Figure 7.1 *For fixed values of model parameters, probabilities underlying the estimation of life expectancies using a one-month grid (black lines) and a one-year grid (grey lines). Upper panel and mid panel: change of transition probabilities given baseline state at age 65. Bottom panel: age-dependent probability to be in state 1.*

Marginal life expectancies are now given by $\widehat{e}_{\bullet 1}(t_1) = 8.61$ and $\widehat{e}_{\bullet 2}(t_1) = 4.75$. The sum of these two quantities is the total life expectancy $\widehat{e}(t_1) = 13.36$.

This computation for $t_1 = 65$ can of course be repeated for $t_1 = 66$, $t_1 = 67$, etc. Figure 7.2 depicts life expectancies for age 65 up to 90. These are marginal expectancies in the sense that state at t_1 is left unspecified and parameters for the logistic model are used; see (7.2). A graph like this nicely shows how total life expectancy subdivides into state-specific life expectancies.

The method in this section is very general. If the multi-state model is defined with Weibull hazards, or is a hybrid model in the sense that various baseline hazard models are included (see Section 4.3), then this does not affect the basic steps in the computation. Once it is clear how to compute the

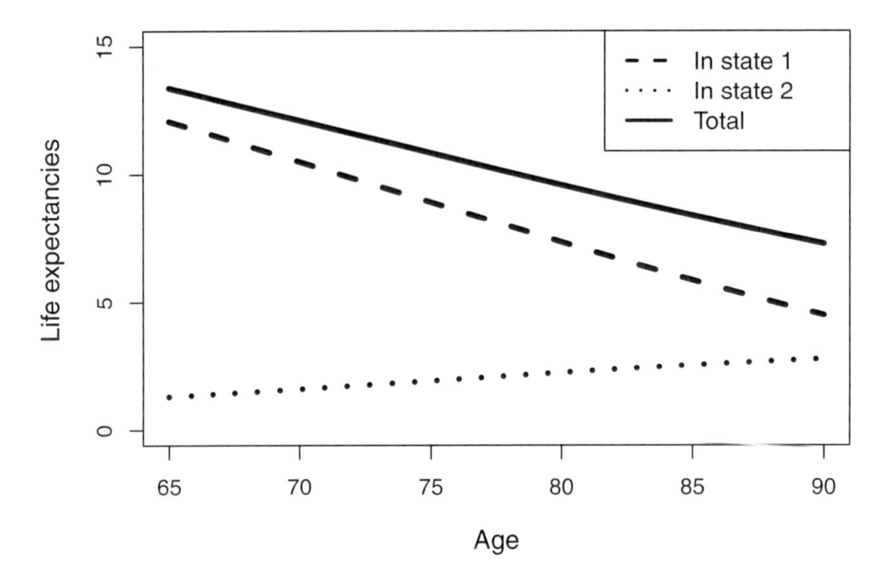

Figure 7.2 *For fixed values of model parameters, marginal life expectancies in years conditional on age.*

transition probability matrix for a given time interval, the steps to compute the life expectancies are the same for all models.

7.5 Computation: Micro-simulation

The basic idea of micro-simulation is that individual trajectories of the multi-state process are simulated many times and that characteristics of the process such as total life expectancy and state-specific life expectancies are derived by analysing the simulated trajectories.

Life expectancies can of course only be estimated by micro-simulation if trajectories are simulated for the whole life span; that is, until the process is absorbed in the dead state. Say the model is a progressive three-state survival model with living states 1 and 2, and dead state 3. The simulation of the multi-state process is based upon a time grid and the simulation of the states at each of the grid points. Assume the grid is given by $u_1 = 0, u_2, ..., u_M$ with time between equal to one year. Given estimates of the parameters for the model, a large number of trajectories through the states are simulated conditional on baseline state 1. If simulated states are denoted by their number, then individual survival in state 1 is derived from the number of 1s in the simulated trajectory.

This is best illustrated by example. Say a simulated trajectory for the three-state model is given by the following series of states and ages:

States:	1	1	1	1	1	2	2	2	2	3
Age:	$u_1 = 65$	66	67	68	69	70	71	72	73	$74 = u_M$

The start state at age 65 is state 1 and the trajectory is absorbed in the dead state after simulating a transition nine times. The simulation here is the simulation of the states at the yearly grid points. It is not a simulation of the transition times. For example, there was a transition from state 1 to state 2 between age 69 and 70, but this is only indicated by a simulated state 2 at age 70. Because the model is progressive, we know in this case that there was not more than one transition. Since we only know the states at the yearly grid points, we assume that transitions take place halfway through the intervals in the grid. From this the years spent in state 1 is computed as $4 + 1/2$ years, and the years spent in state 2 as 4 years.

This computation is basically the counting of the numbers of 1s and 2s in the simulated trajectories, taking into account the start state and the half-year correction. Thus individual survival in state 1 is $(5-1)+1/2 = 4+1/2$ years.

A finer grid is of course possible, in which case the counting of the states has to be adjusted to the grid.

This section illustrates the computation of life expectancies. However, micro-simulation can be used for many other interesting characteristics of an estimated multi-state model. An example for a three-state progressive survival model is the bivariate outcome defined by *age at entry in state* 2 and *duration of stay in state* 2. It is not clear how to derive this outcome using analytic expressions such as the ones used for the life expectancies. However, when using micro-simulation, computation becomes easy: simulate trajectories and monitor both the age at which individuals enter state 2 *and* the time these individuals spent in state 2; see Van den Hout et al. (2014) for an application, and the example in Section 7.6.

When trajectories are simulated for a synthetic population consisting of individuals of the same age and with the same values for background variables, then simulated characteristics of the process will still vary across the individuals due to the random nature of the simulation. For example, even if all individuals start the process in the same state, at the same age, and with the same values of the background variables, simulated death times will still vary. This variation in the simulation leads to uncertainty with respect to the characteristic of interest. This will be called *first-order uncertainty*.

If the micro-simulation takes into account the uncertainty of the estimated model parameters, then this will be a second source of variation in simulated trajectories. With respect to the characteristics of interest, this will be called *second-order uncertainty*.

For the life expectancies, the following introduces micro-simulation formally. Let $\boldsymbol{\theta}$ be the vector with the model parameters. Firstly, values $\boldsymbol{\theta}_b$, $b = 1, ..., B$, are simulated from the distribution of $\boldsymbol{\theta}$. This distribution can be the normal associated with the maximum likelihood estimation (see Section 4.5), but can also be the posterior distribution derived from Bayesian inference. Secondly, conditional on $\boldsymbol{\theta}_b$, a total of C yearly trajectories are simulated and individual state-specific survival times (generically denoted by L_b^c, $c = 1, \cdots, C$) are calculated. Sample mean $\overline{L}_b$ and variance V_b are stored and are used to estimate first-order uncertainty $\mathbb{E}_{\boldsymbol{\theta}}(\text{Var}(L|\boldsymbol{\theta}))$ and second-order uncertainty $\text{Var}_{\boldsymbol{\theta}}(\mathbb{E}(L|\boldsymbol{\theta}))$. The former is the uncertainty caused by variability between individuals in the population, and the latter is what we are after: uncertainty caused by model parameter uncertainty. The sum of these uncertainties is the total variance given by

$$\text{Var}(L) \quad = \quad \mathbb{E}_{\boldsymbol{\theta}}(\text{Var}(L|\boldsymbol{\theta})) + \text{Var}_{\boldsymbol{\theta}}(\mathbb{E}(L|\boldsymbol{\theta})),$$

(cf. Spiegelhalter and Best, 2003). In addition, we use the stored sample means to estimate $\mathbb{E}(L) = \mathbb{E}_{\boldsymbol{\theta}}(\mathbb{E}(L|\boldsymbol{\theta}))$.

O'Hagan et al. (2007) discuss the choice of the number of simulations B and C in a similar nested procedure and present a method to reduce the computational burden in case one is only interested in the mean and the variance of L. In case the goal is the distribution of L, C has to be large. In the example in Section 7.6, it is shown that the method of O'Hagan et al. (2007) can be used to check whether the chosen C was large enough.

Micro-simulation can be implemented in R or in any other software for statistical computing, but the simulation can also be undertaken using the MicMac software (Zinn et al., 2009). This software simulates individual life courses from continuous-time multi-state models, with special emphasis on population projections; see also Willekens (2005) and `www.micmac-projections.org`. MicMac is based on a continuous-time multi-state model with transition intensities that can depend on the age of the individuals and on calendar time. MicMac includes various tools to summarise and display characteristics of the simulated population.

Table 7.1 *Parameter estimates for five-state Model $\mathcal{A}$ for the CFAS data on cognitive function and history of stroke. Estimated standard errors in parentheses, asterisk for a value fix to zero.*

Intercept			t			sex		
$\beta_{12.0}$	-4.481	(0.263)	ξ_{12}	0.135	(0.010)	$\beta_{12.1}$	-0.282	(0.153)
$\beta_{13.0}$	-4.810	(0.275)	ξ_{13}	0.030	(0.017)	$\beta_{13.1}$	0.444	(0.184)
$\beta_{15.0}$	-4.173	(0.137)	ξ_{15}	0.058	(0.010)	$\beta_{15.1}$	0.384	(0.070)
$\beta_{21.0}$	-0.396	(0.252)	ξ_{21}	0^*		$\beta_{21.1}$	0^*	
$\beta_{24.0}$	-2.548	(0.490)	ξ_{24}	ξ_{13}		$\beta_{24.1}$	$\beta_{13.1}$	
$\beta_{25.0}$	-3.123	(0.393)	ξ_{25}	ξ_{15}		$\beta_{25.1}$	$\beta_{15.1}$	
$\beta_{34.0}$	-2.858	(0.413)	ξ_{34}	0.075	(0.017)	$\beta_{34.1}$	0.141	(0.214)
$\beta_{35.0}$	-3.204	(0.370)	ξ_{35}	0.010	(0.013)	$\beta_{35.1}$	$\beta_{15.1}$	
$\beta_{43.0}$	-1.309	(0.320)	ξ_{43}	0^*		$\beta_{43.1}$	0^*	
$\beta_{45.0}$	-0.907	(0.335)	ξ_{45}	ξ_{35}		$\beta_{45.1}$	$\beta_{15.1}$	

7.6 Example: life expectancies in CFAS

To illustrate micro-simulation for the computation of life expectancies, the data from the Cognitive Function and Ageing Study (CFAS) are used; see Section 4.10.1.

The five-state process defined for cognitive function and history of stroke is depicted in Figure 4.7. Various models for this process are specified and fitting using the CFAS data from the Newcastle centre with 2512 individuals; see Section 4.10.2. Here we revisit Model $\mathcal{A}$, where the time dependency is specified by Gompertz baseline hazards. This model is given in (4.17) as

$$q_{rs}(t) = \exp\left(\beta_{rs.0} + \xi_{rs}t + \beta_{rs.1}sex\right).$$

The parameter estimates can be found in Table 7.1, which also provides information on the parameter constraints in Model $\mathcal{A}$. The estimates show that ageing increases the hazard of progression through the states, and that men ($sex = 1$) have a higher hazard for stroke and death than women ($sex = 0$) when controlling for age.

Figure 7.3 depicts state-specific life expectancies for men aged 65, where estimation is based on integration. Simpson's rule is used for integral approximation. The uncertainty in the estimation of the life expectancies as depicted by the kernel densities is derived by using the asymptotic properties of the maximum likelihood estimation of the model parameters. That is, $S = 1000$ model parameter vectors are drawn from the normal distribution associated

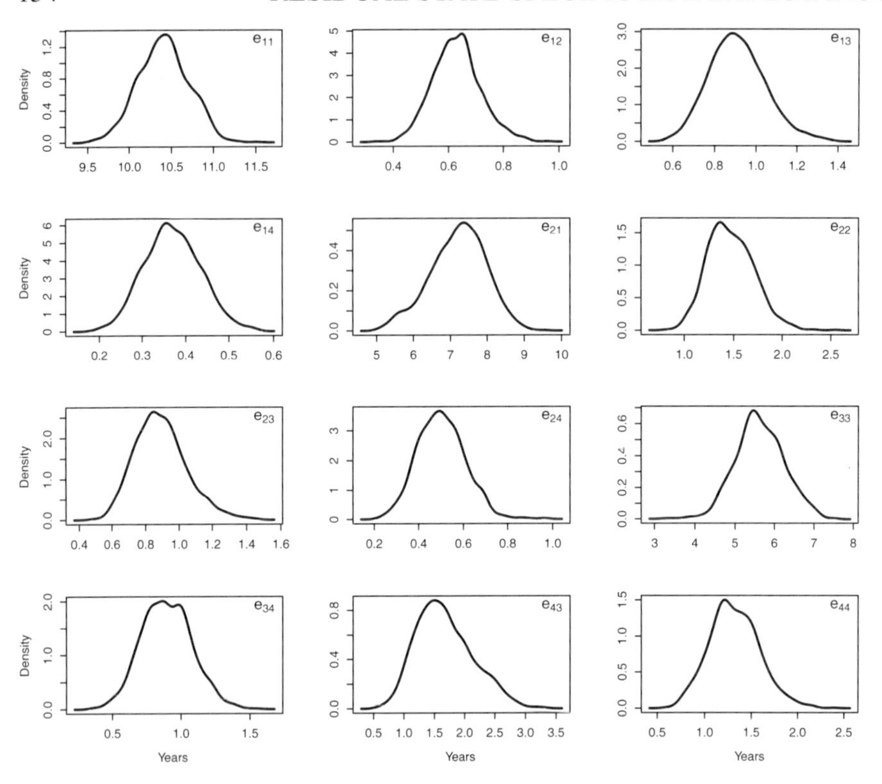

Figure 7.3 *For CFAS, distributions of estimated life expectancies for men aged 65 (derived using $S = 1000$ simulations), where e_{rs} denotes the years in state s given state r at age 65. Using kernel density estimation with Silverman's rule of thumb for the bandwidth.*

with the maximum likelihood estimate. For each of these S vectors, integration is used to derive the life expectancies.

To give an example, life expectancy e_{11} in Figure 7.3 is the number of years spent in state 1 given that a man is in state 1 at age 65. The point estimate is 10.48 and the 95% confidence interval derived from the S iterations is (9.82; 10.96). For women aged 65, the estimate is 12.18 and the interval is (11.59; 12.64). In line with parameter estimates for the effect of *sex*, women are expected to stay longer in state 1 than men of the same age.

Note that the assumption of a symmetric distribution for the uncertainty in the estimation of the model parameters does not imply symmetry for a non-linear transformation of these parameters. Some of the densities in Figure 7.3 are clearly not symmetric; see, for example, e_{43}.

Figure 7.4 depicts state-specific life expectancies for men aged 85. The method for estimation is the same as for age 65. As expected, e_{11} given age

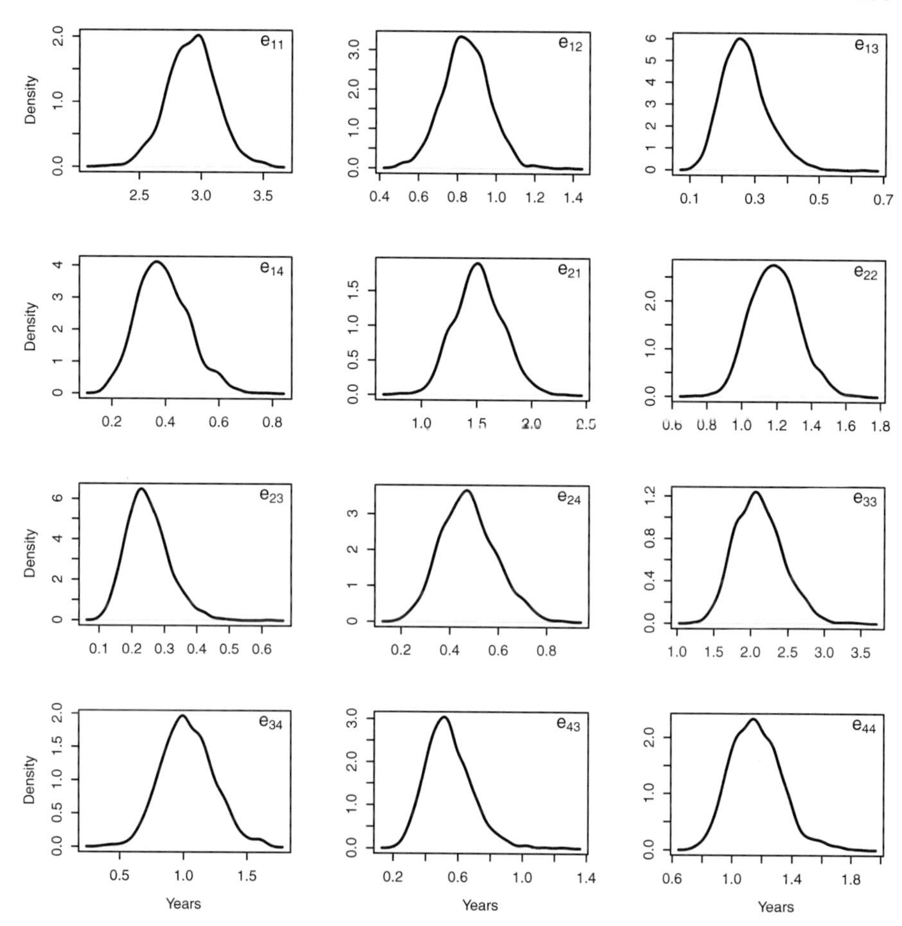

Figure 7.4 *For CFAS, distributions of estimated life expectancies for men aged 85 (derived using* $S = 1000$ *simulations), where* e_{rs} *denotes the years in state s given state r at age 85. Using kernel density estimation with Silverman's rule of thumb for the bandwidth.*

65 is larger than e_{11} given age 85. Note also the variation in e_{21} across the age difference. Although there is no time dependency in the model for the hazard of moving from state 2 to state 1, there is still a strong age effect for e_{21} which is driven by the expected duration of stay in state 1 after a transition to this state from state 2.

Next, we discuss using micro-simulation. The comparison with using integration here is limited to the residual life expectancies e_{11}, e_{12}, e_{13}, and e_{14} for a man in state 1 at age 65. The micro-simulation was undertaken with $B = 200$ and $C = 2000$. The scale for the grid is in years, with grid parameter $h = 1/12$. Table 7.2 presents the numerical comparison. There are small

Table 7.2 *Quantile statistics for estimated residual life expectancies for men aged 65 in state 1. Estimation using integration (with $B = 1000$) and micro-simulation (with $B = 200$ and $C = 2000$).*

	Using integration			Using micro-simulation		
	2.5%	50%	97.5%	2.5%	50%	97.5%
e_{11}	9.819	10.406	10.962	9.902	10.497	11.117
e_{12}	0.473	0.627	0.814	0.461	0.635	0.808
e_{13}	0.667	0.901	1.217	0.620	0.900	1.189
e_{14}	0.248	0.367	0.499	0.240	0.367	0.508

differences between the results of the two computational methods. This is to be expected as both are numerical approximations. Nevertheless, for practical purposes, the differences are negligible.

Using the one-way ANOVA technique proposed by O'Hagan et al. (2007), the micro-simulation results were checked by comparing estimated standard errors. Within-individuals and between-individuals sums of squares are defined by

$$S_{\text{within}} = \sum_{b=1}^{B} \sum_{c=1}^{C} (L_b^c - \overline{L}_b)^2 \quad \text{and} \quad S_{\text{between}} = C \sum_{b=1}^{B} (\overline{L}_b - \overline{L})^2.$$

And the estimator of the second-order uncertainty is

$$v_A = \frac{1}{C} \left(\frac{S_{\text{between}}}{B-1} - \frac{S_{\text{within}}}{B(C-1)} \right),$$

where the definitions of B, C, L_b^c, and $\overline{L}_b$ are given in Section 7.5. As explained, the second-order uncertainty can also be estimated using stored sample means $\overline{L}_b^c$ and variances V_b^c. Let v_M denote this second estimator. The uncertainty is overestimated by v_M if C is not large enough. The bias arises because variability in the $\overline{L}_b$ inflates their variance over and above the variability of the true means that is represented by the second-order uncertainty. Comparison of standard errors $\sqrt{v_A}$ and $\sqrt{v_M}$ is used to assess the bias in v_M and hence the bias in the estimation of L; see also O'Hagan et al. (2007) and Van den Hout and Matthews (2009).

For men aged 65 in state 1, the ANOVA standard errors ($\sqrt{v_A}$) and standard errors directly from the micro-simulation ($\sqrt{v_M}$) are given in Table 7.3. The figures in the table show that for $C = 2000$ there is indeed some consistent overestimation, but that the bias is minimal. Estimated errors for the computation using integration are also presented in Table 7.3, showing good agreement with the results from the micro-simulation.

Table 7.3 *ANOVA standard errors* ($\sqrt{v_A}$) *and standard errors directly from the micro-simulation* ($\sqrt{v_M}$) *for the estimated life expectancies for men aged 65 in state 1. Standard errors (SEs) from the estimation by integration added for comparison.*

	$\sqrt{v_A}$	$\sqrt{v_M}$	SE (using integration)
e_{11}	0.267	0.305	0.300
e_{12}	0.087	0.091	0.086
e_{13}	0.143	0.152	0.136
e_{14}	0.066	0.069	0.065

Because the micro-simulation implies simulating individual trajectories, it is relatively easy to extract additional characteristics of the process. An example of this is the age at entering states 2, 3, and 4 conditional on being in state 1 at age 65. How to write this as a function of model parameters is not directly clear, but it is straightforward to investigate this using simulated trajectories. Using the same micro-simulation as above ($B = 200$, $C = 2000$, and $h = 1/12$), the results for men in state 1 at age 65 are presented by boxplots in Figure 7.5.

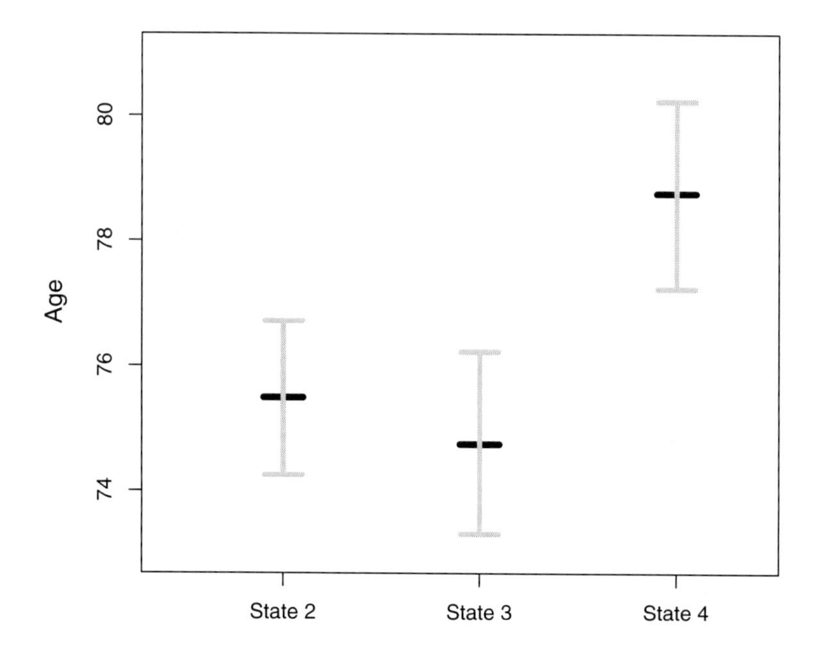

Figure 7.5 *For men, simulated age at entering states 2, 3, and 4 conditional on being in state 1 at age 65. Micro-simulation with* $h = 1/12$, $B = 200$, *and* $C = 2000$. *Black lines for means, grey lines for the range defined by the 2.5% and 97.5% percentiles.*

The results show that the combination of having a history of stroke and being cognitive impaired is associated with an advanced age: point estimate is 78.76 years old, with 95% confidence interval (77.24; 90.23). This information is incomplete, of course, as many men do not enter state 4 at all. Again extracted from the micro-simulation: given state 1 at age 65 the percentages of men entering state 2, 3, and 4 are 0.43 (0.36; 0.51), 0.23 (0.18; 0.30), and 0.27 (0.19; 0.37), respectively (with 95% confidence intervals in parentheses). So the conclusion is that 27% of men in state 1 at aged 65 enters state 4, and that the mean age at entering state 4 conditional on ever entering this state is about 79 years old.

In the above, the inference with respect to mean age at entering a specific living state is presented as a post-hoc analysis of the fitted five-state survival model. As an anonymous reviewer of this manuscript pointed out: the mean age at entry of a specific living state, say state s, can also be inferred by fitting a reduced multi-state model where state s is an absorbing state.

Chapter 8

Further Topics

8.1 Discrete-time model for continuous-time process

Discrete-time stochastic processes as introduced in Section 4.1 can be used to model longitudinal multi-state data. This section discusses discrete-time three-state survival models and compares them with their continuous-time counterparts.

As stated in Section 4.1, transitions in a discrete-time process are timed in steps. In case a step corresponds to a fixed time interval, then the transitions are timed on a uniform grid where the time between the grid points is not part of the model. Only one transition is possible from one grid point to the next. For example, if the process is in state 1 and the only transitions that are possible are $1 \to 2$ and $1 \to 3$, then there are only three possibilities at the next grid point: either the process has stayed in state 1, has moved to state 2, or has moved to state 3.

With regard to longitudinal data analysis, this means that the uniform time grid for a discrete-time model should be suitable for the process at hand. For example, a yearly grid is not suitable when it is quite likely that the process moves from state 1 to state 3 via state 2 within 1 year. Note also that when dead is an absorbing state in the model, death times known on a continuous time scale have to be discretised in order to be taken into account.

An advantage of discrete-time models is that the computation needed for inference is less intensive than the computation for continuous-time models. This is because transition probabilities for discrete-time processes can be modelled directly using generalised linear models; this will be shown below.

In the practice of data analysis, the choice of a discrete-time model versus a continuous-time model will depend on more than one criterion. Are the observations timed on a uniform grid? And if not, can the observation times in the data be approximated by a uniform grid? Is it reasonable to assume a maximum of one transition per fixed time interval? The latter is of specific importance for models with backward transitions. There are no general rules here, but it may help to give some examples.

Diggle et al. (2002, Section 10.3) provide an introduction to discrete-time models with an example of two-state data for respiratory disease among children. The steps of the process are defined by the consecutive observations in the study.

Bacchetti et al. (2010) use a discrete-time model for progression of liver fibrosis following liver transplant. Their model is quite flexible in the sense that it can include misclassification of state, and allows to relax the Markov assumption. For the discrete time scale they use four steps per year. This choice of time scale actually conflicts with a small number of observations in the data. This is solved by changing the corresponding data slightly.

Another example of using discrete-time models is presented by Cai et al. (2010). In an application with interval-censored data on disability and survival in the older population, their model assumes that there are no missing events between two successive observations. Especially, for a model with backward transitions this is restrictive as it implies that when the same state is observed twice in a row, the assumption is that no event has occurred between the two observations. As stated by the authors, when planned observations become less frequent, this assumption can introduce bias in the inference.

See also Bartolucci et al. (2012) for examples of applications in social science and economics.

Even if follow-up data in a longitudinal study are timed on a discrete time scale—say yearly—in practice there will often be variation in times between observations. This variation is ignored if the discrete-time model uses 1-year time units. It is, however, possible to embed continuous-time observations in a discrete-time grid. This is will be illustrated below following the approach in Lièvre et al. (2003), who present a discrete-time model for disability and survival in the older population.

There are also examples with mixed-time formulations; see Dinse (1988) and Lindsey and Ryan (1994). Here the time scale can be continuous for one transition and discrete for another. For the application with data on tumour onset and death, Lindsey and Ryan (1994) prefer the continuous-time specification: slightly more computationally intensive but fewer assumptions and restrictions.

The remainder of this section will define discrete-time progressive three-state models, will explore the models using simulation, and will briefly revisit the Parkinson's disease data introduced in Section 3.7.1.

For a discrete-time progressive three-state survival model, define the one-step transition probabilities for fixed time interval $h > 0$ as

$$\bar{p}_{rs}(t) = \bar{p}_{rs}(t|h) = P(\text{at time } t+h \text{ the state is } s \mid \text{at time } t \text{ the state is } r),$$

for $(r,s) \in \{(1,2),(1,3),(2,3)\}$, and time $t \geq 0$. The transition probabilities not equal to zero can be modelled by

$$\begin{aligned}
\bar{p}_{1s}(t) &= \frac{\exp(\alpha_{1s.0} + \alpha_{1s.1}t)}{1+\exp(\alpha_{12.0}+\alpha_{12.1}t)+\exp(\alpha_{13.0}+\alpha_{13.1}t)} \quad \text{for} \quad s = 2,3 \\
\bar{p}_{11}(t) &= 1 - \bar{p}_{12}(t) - \bar{p}_{13}(t) \\
\bar{p}_{23}(t) &= \frac{\exp(\alpha_{23.0}+\alpha_{23.1}t)}{1+\exp(\alpha_{23.0}+\alpha_{23.1}t)} \\
\bar{p}_{22}(t) &= 1 - \bar{p}_{23}(t) \\
\bar{p}_{33}(t) &= 1. \qquad (8.1)
\end{aligned}$$

These one-step transition probabilities are defined for a discrete-time process where steps are determined by fixed h. However, because the definitions allow for changing t, the model takes time dependency into account. The three equations in (8.1) for leaving or staying in state 1 define a multinominal logistic regression model. The two equations for leaving or staying in state 2 correspond to the common binary logistic regression model. This is the standard way to model the transition probabilities for discrete-time processes (Lièvre et al., 2003; Cai et al., 2010). Additional applications can be found in, for example, Ferrucci et al. (1999) and Cai and Lubitz (2007).

Longitudinal data observed in continuous time can be embedded in the discrete-time grid defined by h. Following Lièvre et al. (2003), for a time interval $(t_1,t_2]$, observation times t_1 and t_2 are rounded to the nearest multiple of h and the duration between the two observations is then expressed in terms of the elementary h-units. Say the rounded times are given by u_1 and u_2, respectively. If the distance between u_1 and u_2 in h-units is m, then the m-step transition probability matrix for $(t_1,t_2]$ is defined as

$$\bar{\mathbf{P}}^{(m)}(t_1,t_2) = \prod_{l=0}^{m-1} \bar{\mathbf{P}}(u_1 + lh),$$

where the (r,s) entry of $\bar{\mathbf{P}}(u_1+lh)$ is $\bar{p}_{rs}(u_1+lh)$.

If the state is right-censored at t_2, then the individual is alive at that time with unknown state. Denote the entries of $\bar{\mathbf{P}}^{(m)}(t_1,t_2)$ by $\bar{p}_{rs}^{(m)}(t_1,t_2)$. We define

$$P(\text{right censored at } t_2 \mid \text{state } r \text{ at } t_1) = \bar{p}_{r1}^{(m)}(t_1,t_2) + \bar{p}_{r2}^{(m)}(t_1,t_2).$$

In case of death at known time t_2, the state just before t_2 is unknown. Hence we define

$$P(\text{death at } t_2 \mid \text{state } r \text{ at } t_1) = \bar{p}_{r1}^{(m)}(t_1,t_2)\bar{p}_{13}(t_2) + \bar{p}_{r2}^{(m)}(t_1,t_2)\bar{p}_{23}(t_2),$$

which implies an unknown state at t_2 followed by a transition to the dead state in the next h-interval. This approach to deal with right censoring and known death times resembles the approach for the continuous-time processes in Section 4.5.

The above defines a transition probability matrix for every observed time interval in the data. Assume that an observed trajectory for individual i consists of observed states y_j at times t_j, for $j = 1,2...,J$. The likelihood contribution for i is given by

$$L_i(\mathbf{y}|\boldsymbol{\alpha}) = \prod_{j=1}^{J-1} P(\text{state } y_{j+1} \text{ at } t_{j+1} \mid \text{state } y_j \text{ at } t_j, \boldsymbol{\alpha}),$$

where $\mathbf{y} = (y_1,...,y_J)$, and $\boldsymbol{\alpha}$ is the vector with the model parameters. The conditional probabilities in the likelihood contribution are derived using m_j-step transition probabilities, with m_j corresponding to the discrete-time grid for interval $(t_j,t_{j+1}]$ for $j = 1,...,J-1$.

The statistical modelling can be extended by including covariate effects in the linear predictors in the generalised linear models (8.1).

Parameter inference can be undertaken by maximisation of the log-likelihood $\sum_{i=1}^{N} \log(L_i(\mathbf{y}_i|\boldsymbol{\alpha}))$. For the maximum likelihood inference a general-purpose optimiser can be used such as `optim` in the R software; see also Section 4.5. An alternative is to use Bayesian inference. Lunn et al. (2013) includes examples of how to use the WinBUGS software to fit discrete-time Markov models.

8.1.1 *A simulation study*

We use simulation to illustrate the discrete-time model and to investigate how well it can approximate a three-state time-dependent progressive process in continuous time. In the simulation study in Section 3.6, trajectories were simulated using a continuous-time progressive model for living states 1 and 2, and dead state 3. The model was defined by the Gompertz baseline hazard. Here we use exactly the same simulation and fit a discrete-time model to every simulated data set. The discrete-time model is defined by (8.1) and estimated by maximum likelihood, where the maximisation is undertaken by using `optim` in R. The time scale is age in years and the fixed time interval h

for the one-step transition probabilities is equal to $1/12$ year, that is, roughly equal to one month.

Because the time-continuous trajectories are simulated using a Gompertz baseline hazard with known parameters, transition probabilities can be calculated for any given time interval $(t_1, t_2]$. This calculation is an approximation because of the numerical approximation of the integral, but we assume that the error of this approximation is negligible and consider the probabilities thus calculated as the true values.

In Section 3.6 there are two designs and here we use Design $\mathcal{B}$ with observations at 0, 4, 8, 9, 10, 11, and 12 years after baseline. For sample size of $N = 200$ and $S = 200$ replications, Table 8.1 presents the results of the simulation for the estimation of the independent entries of $\bar{\mathbf{P}}^{(m)}(t_1 = 15, t_2 = 20)$, where $m = 60$ corresponds to the number of one-month steps to cover $20 - 15 = 5$ years. The results show that Model (8.1) is very good at estimating the transition probabilities for the five-year time interval $(15, 20]$.

Model (8.1) will be called *Gompertz-like*, because the effect of age is modelled in a similar way as in the Gompertz continuous-time model.

Next we investigate the performance of Model (8.1) when the Weibull baseline hazard is used in the simulation. The parameter values in this simulation are $\lambda_{12}, \lambda_{13} = -4$, $\lambda_{23} = -3$, and $\tau_{12} = \tau_{13} = \tau_{23} = 3/2$.

See Table 8.1 for the results. Note that the transition probabilities are different from the ones that were calculated from the Gompertz baseline hazard model—but similar. As expected, using the Gompertz-like discrete-time model causes some bias, most notably for $p_{13}(15, 20)$.

This can be corrected by using a *Weibull-like* discrete-time model. To define this model, first note that $\exp\left(\gamma_0 + \gamma_1 \log(t)\right)$ can be written as $\lambda \tau t^{\tau - 1}$ for $\tau = \gamma_1 + 1$ and $\lambda = \exp(\gamma_0)/\tau$. This leads to the model defined by

$$\begin{aligned}
\bar{p}_{1s}(t) &= \frac{\exp\left(\gamma_{1s.0} + \gamma_{1s.1} \log(t)\right)}{1 + \exp\left(\gamma_{12.0} + \gamma_{12.1} \log(t)\right) + \exp\left(\gamma_{13.0} + \gamma_{13.1} \log(t)\right)} \\
\bar{p}_{23}(t) &= \frac{\exp\left(\gamma_{23.0} + \gamma_{23.1} \log(t)\right)}{1 + \exp\left(\gamma_{23.0} + \gamma_{23.1} \log(t)\right)},
\end{aligned} \tag{8.2}$$

where $s \in \{1, 2\}$. For this model, Table 8.1 shows good simulation results when the data are simulated using a Weibull model.

8.1.2 *Example: Parkinson's disease study revisited*

As an example, consider again the Norway data for Parkinson's disease introduced in Section 3.7.1. The data can be analysed using a three-state progres-

Table 8.1 *Simulation study to investigate the performance of discrete-time models when data are simulated using continuous-time models. Imposed study design with observations at 0, 4, 8, 9, 10, 11, and 12 years. Mean, bias, and root MSE for $S = 200$ replications. Absolute bias less than x is denoted by* $\lceil x \rceil$.

Five-years transition probabilities		Estimation of $\bar{p}_{rs}^{(k)}(15,20)$ Mean	Bias	rMSE
Simulation model Gompertz, analysis model Gompertz-like				
$p_{11}(15,20)$	0.345	0.345	0.001	0.029
$p_{12}(15,20)$	0.153	0.157	0.005	0.023
$p_{13}(15,20)$	0.503	0.497	−0.006	0.030
$p_{22}(15,20)$	0.235	0.235	$\lceil 0.001 \rceil$	0.066
$p_{23}(15,20)$	0.765	0.765	$\lceil 0.001 \rceil$	0.066
Simulation model Weibull, analysis model Gompertz-like				
$p_{11}(15,20)$	0.317	0.336	0.019	0.034
$p_{12}(15,20)$	0.149	0.153	0.004	0.019
$p_{13}(15,20)$	0.534	0.511	−0.023	0.036
$p_{22}(15,20)$	0.210	0.212	0.002	0.050
$p_{23}(15,20)$	0.790	0.788	−0.002	0.050
Simulation model Weibull, analysis model Weibull-like				
$p_{11}(15,20)$	0.317	0.317	$\lceil 0.001 \rceil$	0.029
$p_{12}(15,20)$	0.149	0.152	0.003	0.020
$p_{13}(15,20)$	0.534	0.532	−0.002	0.028
$p_{22}(15,20)$	0.210	0.202	−0.008	0.048
$p_{23}(15,20)$	0.790	0.798	0.008	0.048

sive survival model, where state 1 is defined as being dementia-free, and state 2 is defined as having dementia. The sample size is 233 individuals.

To start, a discrete-time exponential model is fitted using model (8.1) with three intercepts only. For the discrete-time approximation we use $h = 1/12$. Right censoring of states at the end of follow-up, and known death times are dealt with in the likelihood as described above. This intercepts-only model has Akaike's information criterion (AIC) = 2313.

As already demonstrated in Chapter 3, the statistical modelling can be improved by making transition probabilities time-dependent. The chosen time scale t for the Parkinson's disease data is age in years minus 35. For the

Gompertz-like model (8.1) with two α-parameters for each of the three transitions, we obtain AIC = 2278. The Weibull-like model (8.2) with the same number of parameters has AIC = 2265.

In accordance with the analysis using continuous-time baseline models in Section 3.7.2, the Weibull-like model performs better than the Gompertz-like model. However, also for the Weibull-like model the parameters for the transition from state 1 to the dead state 3 have large estimated errors. Restricting the submodel for the transition $1 \to 3$ to be an intercept-only model, we obtain the AIC = 2263. Estimated parameters are

$$\begin{aligned}
\widehat{\gamma}_{12.0} &= -16.319 \; (2.320) & \widehat{\gamma}_{12.1} &= 3.221 \; (0.620) \\
\widehat{\gamma}_{13.0} &= -10.025 \; (0.392) & & \\
\widehat{\gamma}_{23.0} &= -11.454 \; (2.451) & \widehat{\gamma}_{23.1} &= 1.996 \; (0.636) \; ,
\end{aligned}$$

where estimated standard errors are within the parentheses.

One way to interpret the time dependence in the fitted Weibull-like model is that for given ages t_1 and t_2, the ratio $\bar{p}_{12}(t_1)/\bar{p}_{12}(t_2)$ is estimated as $(t_1/t_2)^{\widehat{\gamma}_{12.1}}$. For example, consider two individuals with Parkinson's disease but without dementia. If the first individual is 45 year of age and the other is 40 year of age, then $\bar{p}_{12}(10)/\bar{p}_{12}(5)$ is estimated at 9.324. This implies that the chance for the 45-year-old to get dementia within one year is about nine times the chance for the 40-year-old. Mind that nine times a very small chance is still small. The one-year probability for the 40-year-old is estimated at 0.00023.

8.2 Using cross-sectional data

Throughout this book, multi-state models are fitted to longitudinal data. This section considers options when only cross-sectional data are available.

With longitudinal data there is unit-specific information on the transition process. If the sample units are individuals, then the longitudinal data provide individual-specific information on transitions between the states during the follow-up. The preceding chapters have shown that also with interval-censored transition times, multi-state models can be fitted and many characteristics of the multi-state process can be investigated and predicted. Cross-sectional data need another approach. Here we show how continuous-time models can be used.

There is not a huge literature on using cross-sectional data for transition models. Our approach with a continuous-time model is similar to the approach with discrete-time models in Pelzer et al. (2001), who also provide some pointers to the literature. An alternative is presented in Davis et al. (2002) who derive transition probabilities from modelling and estimating the odds of a transition to another state versus no transition at all. Albarran et al. (2005) discuss the estimation of multi-state survival models using cross-sectional data and annual rates of mortality.

Information on transitions will always be scarce in cross-sectional data. Hawkins and Han (2000) show how cross-sectional data can be used for transition models if partial longitudinal data are available. A different kind of inference on transitions is presented by Van de Kassteele et al. (2012), who discuss a method to estimate net transition probabilities. Net transitions only describe the net inflow or outflow into a state.

8.2.1 Three-state model, no death

The table in Figure 8.1 presents Dutch data on body mass index (BMI). Permission to use the data for this book was kindly given by Jan van de Kassteele. The data are cross-sectional, and men in the age range 15 up to 40 years old are classified in the states 1 = normal weight, 2 = overweight, and 3 = obese; see Van de Kassteele et al. (2012). Instead of longitudinal data on individual men moving between the states, the sample frequencies in the table pertain to different men for each age/state cross-classification. The sample frequencies will be used to illustrate inference for the three-state process as depicted by the diagram in Figure 8.1.

There is an inherent identifiability problem with cross-sectional data for multi-state models. Consider the case of observing the same sample proportions across the yearly age range. It cannot be inferred whether the underlying multi-state process is static in the sense that there are no transitions between the states, or completely symmetric in the sense that a transition $r \to s$ is as likely as $r \leftarrow s$. Mathematically, if the state distribution at age $t-1$ is given by row vector $\boldsymbol{\pi}_{t-1} = (1/3, 1/3, 1/3)$ and the one-year transition probability matrix is given by the 3×3 matrix $\mathbf{P}$, then, for example,

$$\boldsymbol{\pi}_t = \boldsymbol{\pi}_{t-1}\mathbf{P} = \boldsymbol{\pi}_{t-1}$$

for

$$\mathbf{P} = \begin{pmatrix} 1 & 0 & 0 \\ 0 & 1 & 0 \\ 0 & 0 & 1 \end{pmatrix} \quad \textit{and} \quad \mathbf{P} = \begin{pmatrix} 1/2 & 1/2 & 0 \\ 0 & 1/2 & 1/2 \\ 1/2 & 0 & 1/2 \end{pmatrix}.$$

In fact, for every 3×3 double-stochastic matrix $\mathbf{P}$, it follows that $\boldsymbol{\pi}_{t-1}\mathbf{P} = \boldsymbol{\pi}_{t-1}$. A matrix is called double stochastic if each of the rows and each of the

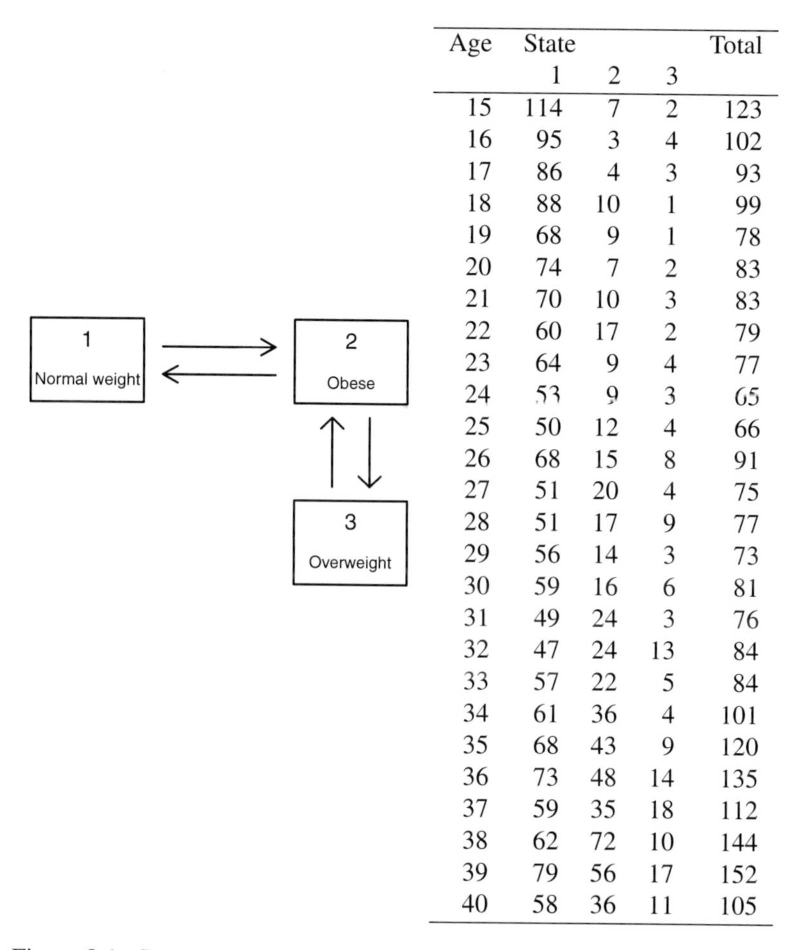

Age	State 1	2	3	Total
15	114	7	2	123
16	95	3	4	102
17	86	4	3	93
18	88	10	1	99
19	68	9	1	78
20	74	7	2	83
21	70	10	3	83
22	60	17	2	79
23	64	9	4	77
24	53	9	3	65
25	50	12	4	66
26	68	15	8	91
27	51	20	4	75
28	51	17	9	77
29	56	14	3	73
30	59	16	6	81
31	49	24	3	76
32	47	24	13	84
33	57	22	5	84
34	61	36	4	101
35	68	43	9	120
36	73	48	14	135
37	59	35	18	112
38	62	72	10	144
39	79	56	17	152
40	58	36	11	105

Figure 8.1 *Cross-sectional data on body mass index for Dutch men from a survey in 2006-2007. Number of observations in the states 1 = normal weight, 2 = overweight, and 3 = obese. Data taken from Van de Kassteele et al. (2012).*

columns sum up to 1. The case with $\boldsymbol{\pi}_{t-1} = (1/3, 1/3, 1/3)$ may not always be relevant to practice, but the issue of identifiability is. For each application, attention is needed with respect to the available data and whether there is sufficient information to estimate the characteristics of the underlying multi-state process. There is no general recipe.

We start with the simple setting in Table 8.2 of a two-state progressive model and cross-sectional data given in the format of age-specific proportions. In this table and in what follows, time scale t is age in years. Define

$$p_{rs}(t-1,t) = P(\text{state } s \text{ at } t \,|\, \text{state } r \text{ at } t-1) \quad \text{for} \quad r,s \in \{1,2\}.$$

Table 8.2: *Cross-sectional data format for a two-state progressive model.*

Age	*Proportion* State 1	State 2
$t-1$	$p_1(t-1)$	$p_2(t-1)$
t	$p_1(t)$	$p_2(t)$

Because the model is progressive, $p_{21}(t-1,t) = 0$ for all t. With the definition $\pi_r(t) = P(\text{state } r \text{ at } t)$, it follows that

$$\pi_1(t) = p_{11}(t-1,t)\pi_1(t-1).$$

Using the age-specific proportions, we get the estimate $\widehat{p}_{11}(t-1,t) = p_1(t)/p_1(t-1)$, and with this we have an estimate of the 1-year transition matrix

$$\begin{aligned}\mathbf{P}(t-1,t) &= \begin{pmatrix} p_{11}(t-1,t) & p_{12}(t-1,t) \\ p_{21}(t-1,t) & p_{22}(t-1,t) \end{pmatrix} \\ &= \begin{pmatrix} p_{11}(t-1,t) & 1-p_{11}(t-1,t) \\ 0 & 1 \end{pmatrix},\end{aligned}$$

for any t for which proportions $p_1(t-1)$ and $p_1(t)$ are available. Yearly transition probabilities for a two-state progressive model can thus be estimated from yearly cross-sectional data.

The above does not hold for the two-state model which includes the transition $2 \rightarrow 1$. For this model $p_{21}(t-1,t) \neq 0$, and the equations for 1 year are

$$\begin{aligned}\pi_1(t) &= p_{11}(t-1,t)\pi_1(t-1) + p_{21}(t-1,t)\pi_2(t-1) \\ \pi_2(t) &= p_{12}(t-1,t)\pi_1(t-1) + p_{22}(t-1,t)\pi_2(t-1),\end{aligned}$$

which is not a set of two linear independent equations due to the sum-to-one restrictions. Depending on the application and the available data, this lack of identification may be dealt with by combining the yearly data and defining a parametric multi-state model.

For the Dutch BMI data in Figure 8.1, Van de Kassteele et al. (2012) present a method to estimate *net transition probabilities*. Net transitions only describe the net inflow or outflow into a state at a certain age, and net transition probabilities are not the same as the transition probabilities which are used throughout this book. As an alternative, the following shows how a parametric three-state model can be fitted using transition probabilities.

Let the time scale t denote age minus 15 years. The hazard models for the transitions between the BMI states are defined as

$$q_{rs}(t) = \exp\left(\beta_{rs} + \xi_{rs} t\right) \quad \text{for} \quad (r,s) \in \{(1,2),(2,1),(2,3),(3,2)\}. \quad (8.3)$$

Estimation of the model parameters is by maximum likelihood, where the likelihood contribution is defined by using the multinomial distribution for frequencies at age t conditional on the proportions at age $t-1$. This approach is similar to using the multinomial distribution for the Bayesian inference in Chapter 6, but instead of using the distribution conditional on individual data as explained in Section 6.5.2, we use the distribution conditional on aggregated data at age t.

Conditional on the row vector $\mathbf{p}_{t-1} = \left(p_1(t-1), p_2(t-1), p_3(t-1)\right)$ with observed proportions for the three states at $t-1$, we assume the distribution of frequencies $\mathbf{Z}_t = \left(Z_1(t), Z_2(t), Z_3(t)\right)$ at t to be multinomially distributed; that is,

$$\mathbf{Z}_t \Big| \, \mathbf{p}_{t-1}, \mathbf{P}(t-1,t), m_t \quad \sim \quad Multinomial\left(\mathbf{p}_{t-1}\mathbf{P}(t-1,t), m_t\right),$$

where $\mathbf{P}(t-1,t)$ is the 3×3 transition matrix for the 1-year interval $(t-1,t]$, and m_t is the number of multinomial trials. For the example with the Dutch data on weight, this implies a probability mass function for each triple of observed frequencies at age 16 up to and including 40, where the number of trials are the sum of the observed frequencies. Let f denote the probability mass function of the multinomial distribution. Assuming independence across the years, the log-likelihood function for the Dutch data is

$$L(\boldsymbol{\theta} | \text{ data }) = \sum_{t=1}^{25} \log\left(f(\mathbf{z}_t | \mathbf{p}_{t-1}, m_t, \boldsymbol{\theta})\right), \quad (8.4)$$

where $\boldsymbol{\theta}$ is the vector with the model parameters and t denotes transformed age. The general-purpose optimiser in R can be used to maximise log-likelihood function (8.4) over the parameter space; see also Section 4.5.

Because information on the transitions is limited in cross-sectional data, it is best to restrict analysis to models which are sparse in the number of parameters and which can be justified by knowledge of the process prior to the data analysis.

For the Dutch data, the first model fitted has only two independent parameters: one for the hazard of moving forward though the states 1, 2, and 3, and one for moving back. Assuming that process of moving forward is different from the process of moving back, this model is as sparse as possible in the current setting. For the specification (8.3), this implies that $\beta_{12} = \beta_{23} \stackrel{\text{d}}{=} \beta_F$, $\beta_{21} = \beta_{32} \stackrel{\text{d}}{=} \beta_B$, and that all ξ-parameters are fixed to zero. This intercept-only model has AIC = 307.

The process of putting on weight is likely to be positively associated with increasing age. Hence the model is extended by adding an age effect for the hazard of moving forward though the states 1, 2, and 3; that is, $\xi_{12} = \xi_{23} \stackrel{\text{d}}{=} \xi_F$ with the rest unchanged. This yields AIC = 297. Parameter ξ_F is estimated at a positive value—as expected.

It may be that the process of loosing on weight is also associated with age. However, extending the model by adding $\xi_{21} = \xi_{32} \stackrel{\text{d}}{=} \xi_B$ does not yield a better fit ($|\widehat{\xi_B}| < 0.001$ and AIC = 299).

The fit of the model with age effect ξ_F only is given by

$$\begin{array}{lllll} q_{12}(t) & = & \exp\left(\beta_F + \xi_F t\right) & \text{with} & \widehat{\beta_F} = -2.81\ (0.32) \\ q_{21}(t) & = & \exp\left(\beta_B\right) & & \widehat{\beta_B} = -0.95\ (0.34) \\ q_{23}(t) & = & \exp\left(\beta_F + \xi_F t\right) & & \\ q_{32}(t) & = & \exp\left(\beta_B\right) & & \widehat{\xi_F} = \ \ 0.06\ (0.01), \end{array}$$

where estimated standard errors are within the brackets.

With estimated model parameters, prediction of the three-state process is possible conditional on prevalence at a specified age. Given the prevalence at age 15, predicted prevalence and observed prevalence are depicted in Figure 8.2. The model with only three parameters is good at capturing the change in the age range 16 up to 40 years old given the prevalence at age 15.

For interpretation of the fitted model, consider the 1-year probability transition matrices for age 25 ($t = 10$) and 35 ($t = 20$):

$$\mathbf{P}(10,11) = \begin{pmatrix} 0.91 & 0.08 & 0.01 \\ 0.29 & 0.64 & 0.07 \\ 0.05 & 0.25 & 0.69 \end{pmatrix} \text{ and } \mathbf{P}(20,21) = \begin{pmatrix} 0.84 & 0.14 & 0.02 \\ 0.27 & 0.60 & 0.13 \\ 0.05 & 0.24 & 0.71 \end{pmatrix},$$

respectively. The matrices illustrate the increase in the risk of overweight with the increase of age.

Equation (8.3) defines a continuous-time multi-state model. For the yearly weight data, a discrete-time model would work similarly well because all the time intervals are fixed to 1 year.

Van de Kassteele et al. (2012) provide yearly weight data for men from birth up to the age of 85. Because of possible association between death and overweight, the data for the older population are not used for the above three-state model. It is also quite likely that the Gompertz hazard as a parametric

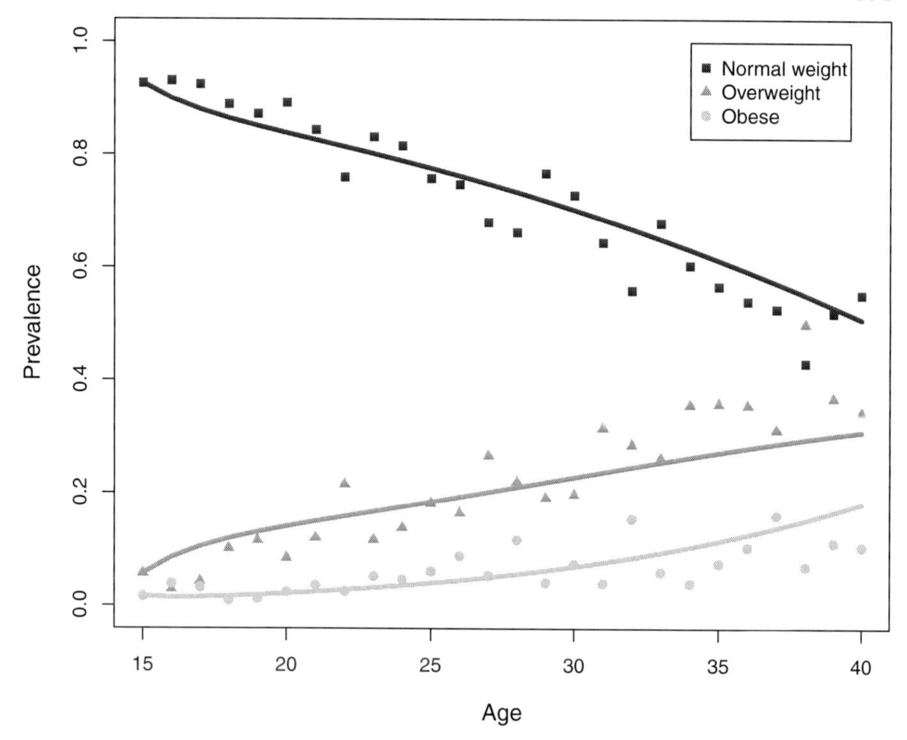

Figure 8.2 *Prevalence and fitted prevalence for Dutch weight data. The fit is conditional on prevalence at age 15.*

shape is too simple to describe the transition-specific risks for the whole age range, which was the second reason to restrict the current analysis to age range 15–40.

8.2.2 Three-state survival model

Given the Dutch BMI data for the three-state model above, if information on yearly survival would also have been provided, would it be possible to fit a four-state survival model for the joint process of survival and gaining weight in the older population? A similar setting for a three-state survival model is discussed in Albarran et al. (2005) for Spanish data on disability. For five age categories, Albarran et al. provide state-specific proportions for two living states, and information on survival which is not state specific.

If the information on survival that is provided is not state specific, identifiability is again an issue. Consider the case of the three-state progressive model with healthy state 1, disease state 2, and dead state 3. Given stable cross-sectional proportions for the living states and survival information which is not state specific, it cannot be inferred whether all deaths are from state 2, or whether deaths occur from both state 1 and state 2. Mathematically, if, for example, the state distribution conditional on being alive at age t is $\boldsymbol{\pi}_t = (1/2, 1/2, 0)$, then

$$\boldsymbol{\pi}_t = \boldsymbol{\pi}_{t-1}\mathbf{P} = (1/4, 1/2, 1/4),$$

for

$$\mathbf{P} = \begin{pmatrix} 1/2 & 1/2 & 0 \\ 0 & 1/2 & 1/2 \\ 0 & 0 & 1 \end{pmatrix} \quad \textit{and} \quad \mathbf{P} = \begin{pmatrix} 1/2 & 2/6 & 1/6 \\ 0 & 2/3 & 1/3 \\ 0 & 0 & 1 \end{pmatrix}.$$

For the following discussion, we assume that there is a consistent change of the cross-sectional prevalence over the age range. This consistency is needed to identify a parametric model. As with the model for the BMI data in the previous section, there is no general recipe to tackle the identification problem.

Assume that cross-sectional data are available for living states 1 and 2 for a range of 1-year age categories. The dead state is state 3. Mortality information

$$p_{nd}(t) = P(\text{no death before age } t \mid \text{alive at age } t-1)$$

is considered to be known for age t in the relevant range. Typically this probability is provided in mortality tables from national statistical offices.

We will start with a non-parametric estimation of the 3×3 transition matrix $\mathbf{P}(t-1,t)$. Define $\pi_1(t) = P(\text{state 1 at } t| \text{ alive at age } t)$. It follows that

$$\begin{aligned} \pi_1(t) &= \frac{P(\text{state 1 at } t| \text{ alive at age } t-1)}{p_{nd}(t)} \\ &= \frac{\pi_1(t-1)p_{11}(t-1,t)}{p_{nd}(t)}. \end{aligned}$$

Define proportion $p_r(t)$ as the sample proportion for state r at age t. Given that the process is progressive with state 3 the dead state and that the mortality information $p_{nd}(t)$ is not state specific, we estimate the unknown entries of $\mathbf{P}(t-1,t)$ by

$$\begin{aligned} \widehat{p}_{11}(t-1,t) &= \frac{p_1(t)p_{nd}(t)}{p_1(t-1)} \\ \widehat{p}_{13}(t-1,t) &= 1 - p_{nd}(t) \\ \widehat{p}_{12}(t-1,t) &= 1 - \widehat{p}_{11}(t-1,t) - \widehat{p}_{13}(t-1,t) \\ \widehat{p}_{23}(t-1,t) &= 1 - p_{nd}(t) \\ \widehat{p}_{22}(t-1,t) &= 1 - \widehat{p}_{22}(t-1,t). \end{aligned}$$

Note that it is not guaranteed that $\widehat{p}_{11}(t-1,t) \in [0,1]$. It might happen that $p_{nd}(t) > p_1(t-1)/p_1(t)$ due to random fluctuation in the data or because there is a structural conflict between the information on mortality and the state-specific frequencies in the data.

As an alternative to the above, we can fit a parametric model. A parametric model can be specified using state-specific hazards for transitions into the dead state. Whether such a model is identified will depend on the available data. One clear advantage of a model is that transition probabilities will be estimated properly in the range $[0,1]$.

The distributional assumption in the model for frequencies $\mathbf{Z}_t = (Z_1(t), Z_2(t), Z_3(t))$ at age t is

$$\mathbf{Z}_t \Big| \, \mathbf{p}_{t-1}, \mathbf{P}(t-1,t), m_t \quad \sim \quad Multinomial\Big(\mathbf{p}_{t-1}\mathbf{P}(t-1,t), m_t\Big), \qquad (8.5)$$

with proportions $\mathbf{p}_{t-1} = \big(p_1(t-1), p_2(t-1), 0\big)$ given by the data for age $t-1$. Random variables $Z_1(t)$ and $Z_2(t)$ have realised values $z_1(t)$ and $z_2(t)$ in the data. Variable $Z_3(t)$ is latent and, as a consequence, m_t is latent, too. However, with $p_{nd}(t)$ given, we can estimate m_t by

$$\widehat{m}_t = \frac{z_1(t) + z_2(t)}{p_{nd}(t)} \qquad \text{for all } t.$$

In the likelihood function, pseudo data $\widehat{z}_3(t) = \widehat{m}_t - z_1(t) - z_2(t)$ can then be used as the realised value of $Z_3(t)$.

We illustrate the approach with simulated data. Let the time scale $t = 0, 1, 2, ..., 20$ denote (transformed) age. The Gompertz hazard model used for the simulation is given by

$$q_{rs}(t) = \exp\left(\beta_{rs} + \xi_{rs}t\right) \quad \text{for} \quad (r,s) \in \{(1,2),(1,3),(2,3)\}, \qquad (8.6)$$

with values $(\beta_{12}, \beta_{13}, \beta_{23}) = (-6, -7, -5)$, and $(\xi_{12}, \xi_{13}, \xi_{23}) = (0.14, 0.04, 0.08)$. The initial distribution at age $t = 0$ is set at $\mathbf{p}(0) = (0.9, 0.1, 0)$. Figure 8.3 presents the three-state model and the simulated cross-sectional data.

Assuming independence across the years, the log-likelihood function for the simulated data is

$$L(\boldsymbol{\theta} | \text{ data }) = \sum_{t=2}^{20} \log\Big(f\big(z_1(t), z_2(t), \widehat{z}_3(t) \,\big|\mathbf{p}_{t-1}, \widehat{m}_t\big)\Big),$$

where $\boldsymbol{\theta}$ is the vector with the model parameters, and f is the probability mass function for the multinomial distribution defined by (8.5).

Model (8.6) was fitted first with the restrictions $\beta_{13} = \beta_{23} \stackrel{\text{d}}{=} \beta_D$, $\xi_{12} = 0$,

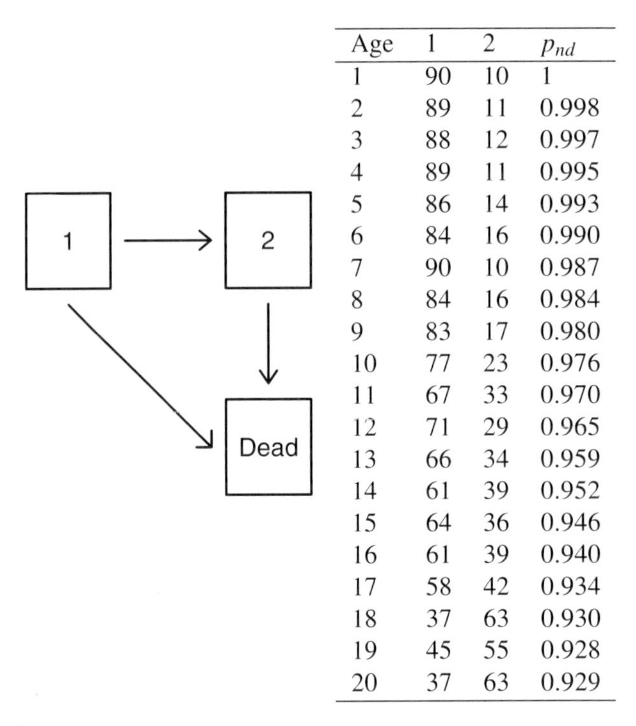

Age	1	2	p_{nd}
1	90	10	1
2	89	11	0.998
3	88	12	0.997
4	89	11	0.995
5	86	14	0.993
6	84	16	0.990
7	90	10	0.987
8	84	16	0.984
9	83	17	0.980
10	77	23	0.976
11	67	33	0.970
12	71	29	0.965
13	66	34	0.959
14	61	39	0.952
15	64	36	0.946
16	61	39	0.940
17	58	42	0.934
18	37	63	0.930
19	45	55	0.928
20	37	63	0.929

Figure 8.3 *Simulated cross-sectional data. Number of observations in the state 1 and 2. Conditional on being alive at age t, p_{nd} is the probability of survival up to age* $t+1$.

and $\xi_{13} = \xi_{23} \stackrel{d}{=} \xi_D$. This model has AIC = 186.4. A better fit was obtained for

$$
\begin{aligned}
q_{12}(t) &= \exp\left(\beta_{12} + \xi_1 t\right) & \text{with} \quad \widehat{\beta}_{12} &= -4.90\ (0.77) \\
q_{13}(t) &= \exp\left(\beta_3 + \xi_1 t\right) & \widehat{\beta}_3 &= -5.20\ (0.39) \\
q_{23}(t) &= \exp\left(\beta_3 + \xi_2 t\right) & \widehat{\xi}_1 &= 0.16\ (0.04) \\
& & \widehat{\xi}_2 &= 0.14\ (0.04).
\end{aligned}
$$

This model has AIC = 183.1. Several sets of starting values were explored, leading to the same estimates. Note that this second model allows for state-specific risk of death in contrast with the non-parametric estimation defined in (8.5).

Figure 8.4 shows estimated transition probabilities. The wiggly line for the non-parametric estimation of p_{12} illustrates the variation in the simulated data. Because the one-year survival probabilities $p_{nd}(t)$ are assumed

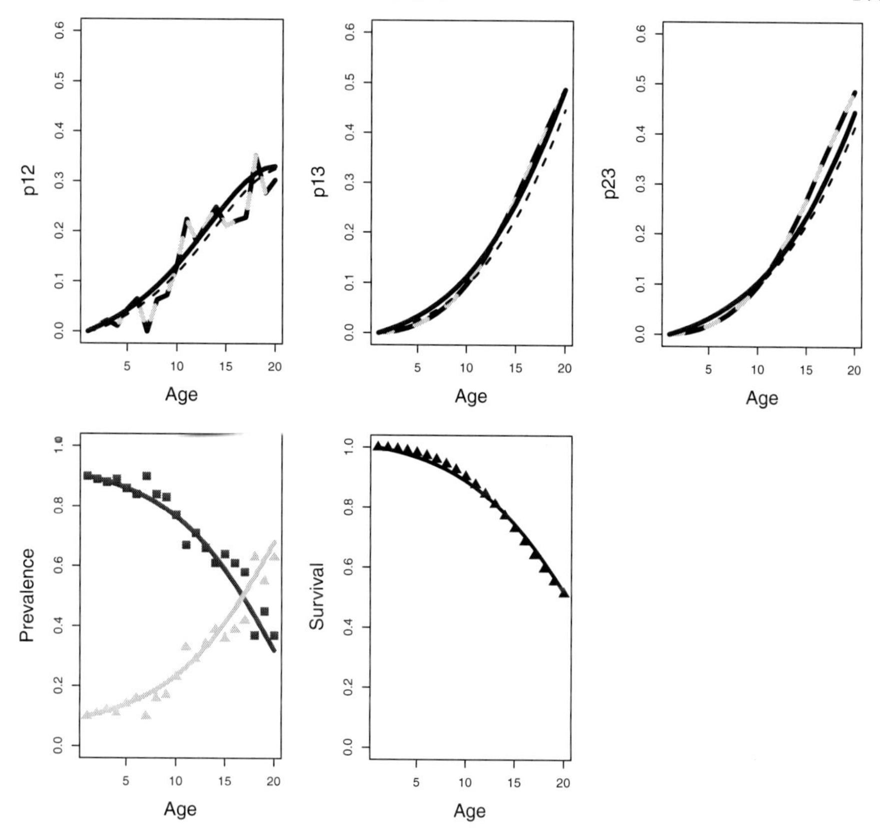

Figure 8.4 *Top panel: estimated transition probabilities for the age range in the simulated three-state survival data. Solid lines for model-based estimation, black/grey lines for non-parametric estimation, dashed lines for true values. Bottom panel: comparison of observed and predicted prevalence and survival. Prediction is model-based and conditional on prevalence at age 0.*

to be known, the values in the data are computed directly from the parameters for the simulation. This explains the smoothness of the curves for the non-parametric estimation of the transition probabilities for the dead state. The parametric estimation of the transition probabilities is close to the non-parametric results. The probabilities p_{13} and p_{23} are estimated at the same values in the non-parametric estimation. For the simulated data, this is not correct. Probabilities p_{13} and p_{23} as estimated by the parametric model are closer to the true values than the non-parametric estimates.

The bottom panels of Figure 8.4 depict the comparison of observed and predicted prevalence and survival. Here the prediction is model-based and conditional on observed prevalence at age 0. The fit is decent.

Given that data were simulated, obtaining a model that fits the data is not a great feat. The general insight is that *if* there is consistent change across the age range of the state prevalence, then multi-state survival models can be investigated using cross-sectional data. However, information will always be limited. In the analysis above, six parameters were used to simulate the data, but the information loss due to the cross-sectional format is such that in this case the underlying unrestricted model (8.6) cannot be identified (huge standard errors for some of the parameter estimates).

The difference between the two sources of the provided data can lead to problems. If the observed state prevalence is limited and noisy but the one-year survival probabilities are smooth in the sense that the information agrees well with a parametric model, then misfit with observed state prevalence is likely to occur in combination with an almost perfect fit to survival. The non-parametric estimation can help to flag this problem.

8.3 Missing state data

Missing data are ubiquitous in applied statistics. Individuals may decline to participate, a measurement device may stop working, a household may move to another country, part of the data may get lost, etc. Especially longitudinal studies are prone to missing data as unforeseen circumstances over time may cause sample units to miss a scheduled assessment or to be lost to follow-up completely.

For multi-state survival models, data can be missing on the state variable and on covariates. We will restrict the current discussion to models for individual trajectories where data are subject to missing values of the state variable. In this setting, two forms of missing values can be distinguished: *intermittently missing values* and missing values due to *lost to follow-up*.

With intermittently missing values there is at least one observation at a later time. Even if the additional data consist of a time for a right-censored living state only, this still provides information since it implies that the individual is not dead at the time of censoring. In contrast, lost to follow-up means that there is no information at all after the last observed state. In the present context of multi-state survival processes, this means that the last observed state was a living state and that there is no follow-up information for the scheduled assessments at a later time. This is also called *premature loss to follow-up*; see, for example, Cook and Lawless (2014).

The data for the multi-state survival models in this book are interval-censored with regard to the transitions between living states. This interval censoring is related to the use of a study design where observations are planned at pre-scheduled times. In this context, intermittently missing values can be seen as interval-censored data with the caveat that the time interval for which no data are available is wider than planned for in the study design.

Using general terminology in missing data analysis, if the probability of observing a state depends on the state itself, then the missing of states is called *non-ignorable* (Little and Rubin, 2002). This implies that the missing-data mechanism and the multi-state process are interlinked and that ignoring the missing data may lead to biased inference.

As an example, say the follow-up is planned every 2 years after baseline. For an individual, state data are given by $(y_1, y_2, \bullet, y_4)$ at times $(t_1, t_2, t_3, t_4) = (0, 2, 4, 6)$, where $\bullet$ denotes a missing value. If the missing of an observed state at t_3 is ignorable, then the method for dealing with the missing value is the same as dealing with interval censoring discussed so far. Note that with a time-dependent hazard model one may consider fine-tuning the piecewise-constant approximation for intervals such as $(t_2, t_4]$. This adjustment, however, does not require new methodology.

The topic of this section is how to deal with non-ignorable missing data. In the example with a missing state value at t_3, this would imply the interval censoring is not independent of the process of interest. Loss to follow-up can also depend on the process of interest. For example, if three living states are defined by employed, unemployed, and retired, then it may be the case that some people stop participating in the study after a transition from unemployed to employed due to time constraints induced by the new job. Not observing these people in the employed state is non-ignorable missing data. Another example is a case where living states are defined by health status. Here non-ignorable missing data are present when there are individuals who stop participating in the study due to deteriorating health.

One way to deal with non-ignorable loss to follow-up is to use inverse probability weights (Hajducek and Lawless, 2012). In what follows, however, we discuss an alternative approach in which the probability of observing a state at time t is modelled conditional on the state at time t. The approach is based upon the work by Cole et al. (2005) who discuss a discrete-time multi-state model for longitudinal data subject to non-ignorable missing values. Their submodel for the ignorable missing states can also be used in combination with a continuous-time multi-state survival model. This will be illustrated for a progressive three-state survival model. The presentation is along the lines of the approach set out in Van den Hout and Matthews (2010).

Given the stochastic process $\{Y_t | t \in (0, \infty)\}$, let R_t denote the observation indicator at time t such that $R_t = 1$ if state Y_t is observed and $R_t = 0$ otherwise.

For an interval $(t_1,t_2]$, state Y_{t_2} and indicator R_{t_2} are modelled jointly by conditioning R_{t_2} on Y_{t_2}. In shorthand notation:

$$P(Y_{t_2},R_{t_2}|Y_{t_1}) = P(R_{t_2}|Y_{t_2},Y_{t_1})P(Y_{t_2}|Y_{t_1}) \tag{8.7}$$

$$= P(R_{t_2}|Y_{t_2})P(Y_{t_2}|Y_{t_1}). \tag{8.8}$$

Using the right-hand conditioning of the indicator R_t in (8.8) on the outcome variables Y_{t_1} and Y_{t_2} is called a selection model in the literature on missing data (Little and Rubin, 2002). Ignoring Y_{t_1} in the distribution of R_{t_2} in (8.8) is a model assumption specific to the current multi-state application. The assumption implies that the reason for missing an observation at time t_2 is dependent on the state at t_2 and not on the state at t_1.

For the observation indicator, two logistic regression models are defined—one model for $R_t = 1$ conditional on $Y_t = 1$, and one model for $R_t = 1$ conditional on $Y_t = 2$. Define the conditional probability that $Y_t = y$ is observed by

$$p_y(t) = P(R_t = 1|Y_t = y, \mathbf{x}_t) \qquad \text{for} \qquad y \in \{1,2\},$$

where $\mathbf{x}_t$ is a covariate vector that includes an intercept. We assume that the probabilities of observing the states can be described by logistic regression models

$$p_y(t) = \frac{\exp(\boldsymbol{\alpha}_y^\top \mathbf{x}_t)}{1+\exp(\boldsymbol{\alpha}_y^\top \mathbf{x}_t)}.$$

The selection model for the bivariate distribution of Y_{t_2} and R_{t_2} for an observed time interval $(t_1,t_2]$ is now given by

$$P(Y_{t_2} = y, R_{t_2} = r|Y_{t_1}, \mathbf{x}_{t_1}) = P(Y_{t_2} = y|Y_{t_1}, \mathbf{x}_{t_1})p_y(t_2)^r \left(1 - p_y(t_2)\right)^{1-r}.$$

Consider an individual i with observation times $(t_1,...,t_J)$. Let $\mathbf{y}^c$ denote the complete-data trajectory $(y_1,...,y_J)$, where the state at t_J is allowed to be right-censored. Let $(r_1,..,r_J)$ be the vector with the observation indicator values. We assume that the baseline state is always observed; that is, $r_1 = 1$. Using the conditional Markov assumption, the contribution of the individual to the complete-data likelihood is given by

$$\begin{aligned} L_i^c(\mathbf{y}^c) &= P(Y_1 = y_1, Y_2 = y_2, ..., Y_J = y_J, R_2 = r_2, ..., R_n = r_n) \\ &= P(Y_2 = y_2, ..., Y_J = y_J, R_2 = r_2, ..., R_n = r_n|Y_1 = y_1)P(Y_1 = y_1) \\ &= L_2^c \times \cdots \times L_J^c \times P(Y_1 = y_1), \end{aligned}$$

where the conditioning of the covariates is suppressed in the notation. The contributions L_j^c for $j \in \{2,...,J\}$ are defined as follows. If the state at t_j, $j \in \{2,..,J\}$, is 1 or 2, then

$$L_j^c = P(Y_j = y_j | Y_{j-1} = y_{j-1}) p_{y_j}(t_j)^{r_j} \left(1 - p_{y_j}(t_j)\right)^{1-r_j}.$$

If the state at t_J is 3, that is, the dead state, then

$$L_J^c = P(Y_J = 1 | Y_{J-1} = y_{J-1}) q_{13}(t_{J-1}) + P(Y_{t_J} = 2 | Y_{J-1} = y_{J-1}) q_{23}(t_J).$$

If the state is right-censored at t_J, we assume that the individual is alive but with unknown state and define

$$L_J^c = P(Y_J = 1 | Y_{J-1} = y_{J-1}) + P(Y_J = 2 | Y_{J-1} = y_{J-1})$$

The likelihood contribution $L_i(\mathbf{y})$ of individual i is now defined by summing over all possible missing values:

$$L_i(\boldsymbol{\theta} | \mathbf{y}, \mathbf{x}) = P(Y_1 = y_1) \sum_{\mathbf{y}^c \in \Omega(\mathbf{y})} L_i^c(\mathbf{y}^c),$$

where $\boldsymbol{\theta}$ is the vector with the model parameters, $\Omega(\mathbf{y})$ is the set with all the trajectories where missing states are replaced by feasible latent states, and covariate vector $\mathbf{x}$ contains the covariates for the hazard models and for the missing data model. For a progressive three-state survival model, only patterns with monotone increase are possible. For example, if $\mathbf{y} = (1, \bullet, \bullet, 2, 3)$, then $\Omega(\mathbf{y}) = \{(1,1,1,2,3), (1,1,2,2,3), (1,2,2,2,3)\}$.

As in Section 4.5, assuming that the parameters for the initial state and the subsequent transitions are distinct, we can ignore the initial state distribution $P(Y_1 = y_1)$ when estimating the parameter of the multi-state model.

The formulation of the likelihood contributions resembles the expressions in Cole et al. (2005) where a discrete-time model is specified for data without exact death times and right censoring. A continuous-time model is defined by transition intensities, but for maximum likelihood estimation, the intensities are translated to transition probabilities; hence the similarity in the likelihoods, even though the models are quite different.

Given a three-state survival model without recovery, the value of a missing living state after a previously observed state 2 is always state 2. For example, if $\mathbf{y} = (y_1, y_2, y_3, y_4) = c(1, 2, \bullet, 3)$, then we know that $y_3 = 2$. The fact that state y_3 is not observed is used in the estimation of the logistic regression model for the probability of observing state 2. Because we do not want to loose information on the missing data mechanism, these kind of missing states should not be imputed before the data are analysed even though we know the unobserved states.

With the likelihood function specified, a general-purpose optimiser can be used to estimate model parameters. This can be computationally intensive when there are many individuals with a large set $\Omega(\mathbf{y})$.

For a progressive three-state survival model, Van den Hout and Matthews (2010) ran a simulation study to investigate the performance of the selection model above, and to investigate bias when *(i)* the missing data are not taken into account as ignorable missing data, and *(ii)* the missing data are not taken into account at all. Both *(i)* and *(ii)* lead to bias results. The multi-state data analysis on stroke and survival in Van den Hout and Matthews (2010) shows that whether or not missing data are taken into account can have a substantial effect on long-term prediction.

8.4 Modelling the first observed state

This section discusses left truncation of data for multi-state survival models. A typical example of left truncation in this setting is using age as the time scale for analysing longitudinal data where individuals are included in the baseline conditional on having reached a certain age.

One way to deal with the left truncation is to condition on the first observed state. This is the approach for the individual-specific likelihood contribution in Section 4.5. It implies that only transition probabilities $P(Y_j = y_j | Y_{j-1} = y_{j-1}, \boldsymbol{\theta}, \mathbf{x})$ have to be modelled, where $\boldsymbol{\theta}$ is the vector with the model parameters, observations of state $y_1, y_2, \ldots, y_J$ are timed at $t_1, t_2, \ldots t_J$, respectively, and covariate information is collected in $\mathbf{x}$. This approach can also be used when entry to the study is restricted to individuals in a subset of the living states; see, for example, the four-state survival model in Joly et al. (2009) where individuals are followed up only when they are in the healthy state at the initial observation.

An alternative is to define the contribution of individual i to the likelihood function by

$$L_i(\boldsymbol{\theta}|\mathbf{y},\mathbf{x}) = P(Y_J = y_J, \ldots, Y_2 = y_2, Y_1 = y_1 | Y_1 \neq D, \boldsymbol{\theta}, \mathbf{x}).$$

The condition $Y_1 \neq D$, where D denotes the dead state, implies that the individual is alive at the baseline of the study. This is a straightforward generalisation of left truncation as introduced for the survival model in Section 2.1.

Using the conditional Markov assumption, the contribution can be written as

$$L_i(\boldsymbol{\theta}|\mathbf{y},\mathbf{x}) = P(Y_1 = y_1|Y_1 \neq D, \boldsymbol{\theta}, \mathbf{x}) \times \left(\prod_{j=2}^{J-1} P(Y_j = y_j|Y_{j-1} = y_{j-1}, \boldsymbol{\theta}, \mathbf{x})\right) C(y_J|y_{J-1}, \boldsymbol{\theta}, \mathbf{x}), \tag{8.9}$$

where $C(y_J|\ldots)$ is defined in Section 4.5. Ignoring the conditioning on $\boldsymbol{\theta}$ and $\mathbf{x}$ in the notation, we have

$$P(Y_1 = y_1|Y_1 \neq D) = \frac{\sum_{s_0 \neq D} P(Y_1 = y_1|Y_0 = s_0)P(Y_0 = s_0)}{\sum_{s_0 \neq D} P(Y_1 \neq D|Y_0 = s_0)P(Y_0 = s_0)}, \tag{8.10}$$

where Y_0 refers to state at $t = 0$. The distribution of the state at $t = 0$ is not described by the multi-state model as specified previously. Hence, specification (8.10) needs additional modelling of the state prevalence at $t = 0$.

The above assumes that $t = 0$ is relevant to the study at hand. When all individuals are in state s_0 at $t = 0$, then (8.10) simplifies to

$$P(Y_1 = y_1|Y_1 \neq D) = \frac{P(Y_1 = y_1|Y_0 = s_0)}{P(Y_1 \neq D|Y_0 = s_0)}.$$

When left truncation is present, the choice of time scale is important. Consider the cardiac allograft vasculopathy (CAV) study where all individuals are in the first state at the start of the follow-up; see Figure 1.1. If the time scale is years since transplant, then $t_1 = 0$, and the baseline state $y_1 = s_0 = 1$ for all individuals. Hence $P(Y_1 = y_1|Y_1 \neq D) = 1$ in (8.10) and can thus be ignored in the data analysis. If the time scale for modelling the CAV data is age, then dealing with the left truncation as set out above is not possible: the model for the multi-state process after transplant does not hold for individual age before the transplant.

Consider as a second example a study of cognitive function and survival in the older population, where baseline age is 65 years for all individuals. Even though left truncation is present in the sense that individuals have to reach age 65 to enter the study, there are no data on the process that underlie the truncation (the multi-state process before age 65). Since all individuals in the study are alive at baseline age 65, the model assumes that $P(Y_1 = y_1|Y_1 \neq D)$ is equal to $P(Y_1 = y_1)$. Of course, given a multi-state survival model for cognitive function, not all individuals in the study will be in the same living state at baseline age 65. This can be taken into account by modelling the distribution of the first observed state $P(Y_1 = y_1)$ using a multinomial logistic regression model.

As a third example, assume that the above study of cognitive function and survival is designed in such a way that baseline age varies from 65 years and older. If the likelihood-function contribution is defined by (8.9), then for those individuals who are older than 65 at baseline, data on survival up to their baseline age are available from younger individuals in the study. The problem here is that the state occupancy before the first observed state at baseline is unknown. A possible way to deal with this is, firstly, setting $t = 0$ equal to an age at which all individuals in the population are assumed to be in the same state—typically state 1 in an illness-death model, and, secondly, assuming that the model for the multi-state process holds for $t = 0$ onward. This approach with respect to the time scale can also be found in the literature on semi-Markov models; see, for example, Kapetanakis et al. (2013).

8.5 Misclassification of states

When measuring a stochastic process using a discrete set of states, measurement error implies that a latent true state r is observed as state $r^* \neq r$. This is called misclassification of state. In this setting, we will denote the observed stochastic process by $\{Y_t^* | t \in (0, \infty)\}$ with state space $\mathcal{S}^*$, and the latent process by $\{Y_t | t \in (0, \infty)\}$ with state space $\mathcal{S}$.

If the Markov assumption is used for the latent stochastic process, then misclassification models are called hidden Markov models. Especially in biostatistics, misclassification of state is quite common; see, for example, the continuous-time processes in Satten and Longini (1996), Bureau et al. (2003), and Jackson et al. (2003).

Misclassification can occur when a marker for a condition or disease is categorised in a series of living states using threshold values. Van den Hout et al. (2014) discuss an example with data on cognitive function and survival in the older population. Their model for the latent multi-state process does not allow transitions from a cognitive-impaired state back to a non-impaired state—backward transitions are explained by misclassification. Including misclassification is seen as a way to smooth the data; the measurement of cognition over time may result in a backward transition now and then, but the underlying latent cognitive process is assumed to be stable or decreasing.

If misclassification of state is present in longitudinal data, a model for the latent process Y_t has to be combined with a model for the misclassification. To introduce the modelling, the time-homogeneous misclassification probability for observed state r^* and latent state r is defined as

$$c_{rr^*} = P(Y_t^* = r^* | Y_t = r). \tag{8.11}$$

Let us assume that $\mathcal{S}^* = \mathcal{S} = \{1, 2, ..., D\}$. This implies that the misclassification probabilities c_{rr^*} define a $D \times D$ matrix, say $\mathbf{C}$, with the rows summing up to one. If a state r is always classified correctly, then, of course, $c_{rr} = 1$ and $c_{rr^*} = 0 \; \forall r^* \neq r$.

Commonly, misclassification probabilities are modelled using the logit link. For the intercept-only model for time-homogeneous probabilities this implies $c_{rr^*} = \exp(\alpha_{rr^*})/(1 + \exp(\alpha_{rr^*}))$ for $\alpha_{rr^*} \in \mathbb{R}$. This basic model can be extended by adding covariates, or by including a dependence on time.

A hidden Markov model can be estimated using maximum likelihood. Assume that an individual i has observed states $\mathbf{y}^* = (y_1^*, ..., y_J^*)$ at times $(t_1, ..., t_J)$, and time-independent covariate vector $\mathbf{x}$. The contribution of this individual to the likelihood function is

$$\begin{aligned} &L_i(\boldsymbol{\theta}|\mathbf{y}^*, \mathbf{x}) \\ &\quad = P(Y_J^* = y_J^*, ..., Y_1^* = y_1^*) \\ &\quad = \sum_{(y_1,...,y_J) \in \Omega_J} P(Y_J^* = y_J^*, ..., Y_1^* = y_1^* | Y_J = y_J, ..., Y_1 = y_1) P(Y_J = y_J, ..., Y_1 = y_1), \end{aligned}$$

where $\boldsymbol{\theta}$ is the vector with all model parameters, and Ω_J is the set with all possible paths of latent states at times $(t_1, ..., t_J)$. In this notation the dependence on covariates and model parameters is ignored at the right-hand side.

The latent process Y_j is assumed to be Markovian conditional on time t_j and covariates at time t_j. With respect to the misclassification, for every pair of observed states Y_j^* and Y_k^*, $j \neq k$, we assume

$$P(Y_j^* = y_j^*, Y_k^* = y_k^* | Y_j = y_j, Y_k = y_k) = P(Y_j^* = y_j^* | Y_j = y_j) P(Y_k^* = y_k^* | Y_k = y_k).$$

This means that misclassification at time t_j is independent of latent states at other times and is independent of misclassification at other times. We obtain in shortened notation

$$\begin{aligned} &L_i(\boldsymbol{\theta}|\mathbf{y}^*, \mathbf{x}) \\ &\quad = \sum_{y_1} P(Y_1^*|Y_1) P(Y_1) \sum_{y_2} P(Y_2^*|Y_2) P(Y_2|Y_1) \times \cdots \times \sum_{y_J} P(Y_J^*|Y_J) P(Y_J|Y_{J-1}); \end{aligned} \tag{8.12}$$

see Satten and Longini (1996). Probabilities $P(Y_j^*|Y_j)$ are the classification probabilities (8.11) and $P(Y_j|Y_{j-1})$ are the transition probabilities for the latent multi-state process. If unknown, the distribution of the latent first state $P(Y_1) = P(Y_1 = y_1)$ can be modelled by multinomial logistic regression.

As shown in Satten and Longini (1996), matrices can be used to describe the likelihood function contribution (8.12). Let $\mathbf{f}$ be the $1 \times D$ row vector with rth element equal to $P(Y_1^* = y_1^* | Y_1 = r)P(Y_1 = r)$. For $j = 2, ..., D-1$, let $\mathbf{T}_j$ be the $D \times D$ matrix with (r, s) entry

$$P(Y_j^* = y_j^* | Y_j = s)P(Y_j = s | Y_{j-1} = r).$$

The definition of $\mathbf{T}_D$ depends on the last observation y_J. In case of death, that is, $Y_J^* = D$, the (r, s) entry of $\mathbf{T}_D$ is

$$P(X_D = s | Y_{D-1} = r)q_{sD}(t_D).$$

So we assume an unknown latent state s at time t_D and then an instant death. In case of right censoring, it is known that the individual is still alive at time t_D but a living state is not observed. In this case, the (r, s) entry of $\mathbf{T}_D$ is

$$P(Y_D = s | Y_{D-1} = r)\,\mathbb{1}_{[s \in \{1,2,..,D-1\}]},$$

where $\mathbb{1}_{[A]}$ is equal to 1 if A is true and zero otherwise. The individual's likelihood contribution is

$$L_i(\boldsymbol{\theta} | \mathbf{y}^*, \mathbf{x}) = \mathbf{f}\mathbf{T}_2\mathbf{T}_3 \times \cdots \times \mathbf{T}_D\mathbf{1},$$

where $\mathbf{1}$ is the all-one $D \times 1$ vector.

As stated in Bureau et al. (2003), identifiability can be a problem in a hidden Markov model. Van den Hout et al. (2009) present the following example of a three-state model which is not identified in the misclassification probabilities. Let state 3 be the dead state. Say there are two individuals with observation times t_1 and t_2. Observed states of the first individual are $y_1^* = 1$ and $y_2^* = 3$. Observed states of the second individual are $y_1^* = 2$ and $y_2^* = 3$. Assume that the individuals have the same covariate values. The likelihood function is given by

$$\begin{aligned} L(\boldsymbol{\theta} | \mathbf{y}^*, \mathbf{x}) = \; & \big(P(Y_1^* = 1 | Y_1 = 1)P(Y_1 = 1)P(Y_2 = 3 | Y_1 = 1) \\ & \quad + P(Y_1^* = 1 | Y_1 = 2)P(Y_1 = 2)P(Y_2 = 3 | Y_1 = 2)\big) \\ & \times \big(P(Y_1^* = 2 | Y_1 = 1)P(Y_1 = 1)P(Y_2 = 3 | Y_1 = 1) \\ & \quad + P(Y_1^* = 2 | Y_1 = 2)P(Y_1 = 2)P(Y_2 = 3 | Y_1 = 2)\big), \end{aligned}$$

where we used the fact that $P(Y_2^* = 3 | Y_2 = 3) = 1$. Rewriting the likelihood using the parameters, we obtain

$$L(\boldsymbol{\theta} | \mathbf{y}^*, \mathbf{x}) = \Big((1 - c_{12})g_1(\boldsymbol{\theta}) + c_{21}g_2(\boldsymbol{\theta})\Big)\Big(c_{12}g_1(\boldsymbol{\theta}) + (1 - c_{21})g_2(\boldsymbol{\theta})\Big),$$

where g_1 and g_2 are functions of those model parameters in $\boldsymbol{\theta}$ that define the Markov model and the model for the first state. As an example, values $(c_{12}, c_{21}) = (1/4,\ 1/5)$ and $(c_{12}, c_{21}) = (3/4,\ 4/5)$ yield the same likelihood function value when the other model parameters stay fixed.

Especially in small samples, one might end up in a local maximum where the diagonal of estimated $\mathbf{C}$ does not dominate. However, if one is confident that the true state of affairs is described by a matrix $\mathbf{C}$ with a dominating diagonal, a rerun with different starting values can be undertaken.

8.5.1 *Example: CAV study revisited*

The CAV study is introduced in Section 1.3.1. It is a follow-up study with individual histories of 622 heart transplant recipients. Four CAV states are defined: state 1 = no CAV, state 2 = mild/moderate CAV, state 3 = severe CAV, and state 4 = dead; see also Figure 1.1.

CAV is assumed to be a progressive process where recovery is not possible. Nevertheless, backward transitions from states 2 and 3 are observed in the data; see Table 1.1. In Section 1.3.2, a progressive multi-state model is fitted to data where the state at time t is defined as the highest CAV state $\in \{1,2,3,4\}$ observed in $(0,t]$. Time scale t is years since transplant. In what follows, we adopt the approach in Sharples et al. (2003) and fit multi-state models for the latent progressive process where backward transitions $2 \rightarrow 1$ and $3 \rightarrow 2$ are dealt with by including misclassification of states 1, 2, and 3.

The analysis start with Model A, which is described in Jackson (2011), and is given by

$$\begin{aligned} q_{rs}(t) &= \exp(\beta_{rs.0}) & \text{for} \quad (r,s) \in \{(1,2),(1,4),(2,3),(2,4),(3,4)\} \\ c_{rr^*} &= \frac{\exp(\alpha_{rr^*})}{1+\exp(\alpha_{rr^*})} & \text{for} \quad (r,r^*) \in \{(1,2),(2,1),(2,3),(3,2)\}. \end{aligned}$$

The CAV data are provided in the R-package msm, and Jackson (2011) presents the R code explicitly for fitting Model A using msm. In msm, the maximisation of the log-likelihood function is undertaken by using the general-purpose optimiser `optim` in R. Alternatively, one can implement the log-likelihood function in R, and use `optim` directly. There is a wide range of options in msm to extend basic Model A. Likewise, once the log-likelihood function in R is implemented, coding the extended models is relatively easy.

Model A has nine parameters and AIC = 3928.1. In line with the analysis in Section 1.3.2, the statistical modelling is extended for transitions $1 \rightarrow 2$ and $1 \rightarrow 4$ with Gompertz time dependency and covariates *b.age* (baseline age at transplant) and $d.age$ (donor age). This defines Model B, for which the hazard specifications are

$$\begin{aligned} q_{rs}(t) &= \exp(\beta_{rs.0} + \xi_{rs}t + \beta_{rs.1}b.age + \beta_{rs.2}d.age) \\ & \qquad\qquad \text{for} \quad (r,s) \in \{(1,2),(1,4)\} \\ q_{rs}(t) &= \exp(\beta_{rs.0}) \quad \text{for} \quad (r,s) \in \{(2,3),(2,4),(3,4)\}. \end{aligned}$$

Model B has 18 parameters and AIC = 3872.6. With the extra information which is included in Model B, it is not a surprise that this model performs better than intercept-only Model A.

Just to illustrate the scope for statistical modelling, two additional models are defined. For Model C, the logit link for the misclassification probabilities in Model B is replaced by a probit link; that is, $c_{rr^*} = \Phi(\alpha_{rr^*})$ with Φ the cumulative distribution function of the standard normal. In practice, changing from a logit link to a probit link will, in most cases, induce the same inference. Nevertheless, it is nice to be able to check this for the data at hand. Rounded to one decimal, Model C has the same AIC as Model B.

Parameter estimates for Model B are presented in Table 8.3. Looking at the estimated misclassification parameters, it seems opportune to define Model D as a restricted version of Model B by imposing $\alpha_{23} = \alpha_{32}$. This restriction implies that it is as likely to misclassify latent state 2 as state 3 as it is to misclassify a latent state 3 as state 2. Model D has 17 parameters and AIC = 3872.0. Even though the AIC is slightly lower than the AIC of Model B, the difference is minor. Because state 3 will also include the very severe CAV cases for which misclassification is unlikely, we prefer Model B over Model D.

It is interesting to compare the estimated parameters in Model B with the estimates reported in Table 1.2 for the process defined by observed highest CAV state. Because the two tables are derived using two different data sets, any comparison is tentative. Nevertheless, for the effects of time since transplant, age at transplant, and donor age, we see that estimated effects which have relatively small standard errors are similar in both tables.

A more informative comparison can be made by looking at derived transition probabilities. As in Section 1.3.2, assume we are interested in the first year after the transplant given median age at transplant 49.5 years, and median donor age 28.5. Conditional on these median values, the probability matrix for

Table 8.3 *Parameter estimates for four-state Model B for the CAV data. Estimated standard errors in parentheses. Time scale t is years since transplant, b.age is age at transplant, and d.age is age of the donor.*

Intercept		t		$d.age$	
$\beta_{12.0}$	−3.601 (0.365)	$\beta_{12.1}$	0.119 (0.028)	$\beta_{12.3}$	0.029 (0.007)
$\beta_{14.0}$	−6.536 (0.865)	$\beta_{14.1}$	−0.014 (0.079)	$\beta_{14.3}$	0.029 (0.010)
$\beta_{23.0}$	−1.066 (0.375)			$\beta_{23.3}$	−0.012 (0.011)
$\beta_{24.0}$	−1.571 (1.120)	*b.age*		$\beta_{24.3}$	−0.048 (0.040)
$\beta_{34.0}$	−0.669 (0.369)	$\beta_{12.2}$	0.003 (0.007)	$\beta_{34.3}$	−0.017 (0.012)
		$\beta_{14.2}$	0.050 (0.016)		
Misclassification parameters:					
$\alpha_{12.0}$	−3.752 (0.289)	α_{23}	−2.720 (0.285)		
$\alpha_{21.0}$	−1.470 (0.240)	α_{32}	−2.117 (0.383)		

the first year is given by

$$\widehat{\mathbf{P}}(t_1 = 0, t_2 = 1) = \begin{pmatrix} 0.897 & 0.057 & 0.007 & 0.039 \\ 0 & 0.740 & 0.181 & 0.079 \\ 0 & 0 & 0.730 & 0.270 \\ 0 & 0 & 0 & 1 \end{pmatrix}.$$

Comparing this matrix with the matrix derived from the model for observed history of CAV, we see that the Model B with misclassification predicts faster progression to a higher state once in state 2 or state 3. For the first three elements of the diagonal, the 95% confidence intervals are (0.877,0.912), (0.679,0.780), and (0.672,0.782), respectively.

The α-parameters for the misclassification in Model B are on the logit scale. The estimated misclassification matrix derived using the α-parameters is given by

$$\widehat{\mathbf{C}} = \begin{pmatrix} 0.98\ (0.96, 0.99) & 0.02\ (0.01, 0.04) & 0 & 0 \\ 0.19\ (0.11, 0.29) & 0.75\ (0.60, 0.86) & 0.06\ (0.04, 0.10) & 0 \\ 0 & 0.11\ (0.05, 0.20) & 0.89\ (0.80, 0.95) & 0 \\ 0 & 0 & 0 & 1 \end{pmatrix},$$

where 95% confidence intervals are given within the parentheses. According to Model B, there is a substantial chance that a latent state 2 is misclassified as either state 1 or state 3.

8.5.2 *Extending the misclassification model*

The example above with the CAV data illustrates the basic approach for dealing with misclassification in a time-continuous multi-state survival model. In the CAV study, the time scale is years since baseline, and we assume that at baseline all individuals are in latent state 1. Hence, for the CAV study there is no need to model the first latent state. However, many other longitudinal studies will require a model for the first latent state when misclassification is included.

In the example given in Section 8.5 with data on cognitive function and survival in the older population, the time scale in the model for the latent state is age in years. In this case, both baseline age and latent state at baseline vary across individuals. Multinomial logistic regression can be used to model the first latent state, where, typically, one wants to include baseline age as one of the covariates.

When misclassification of states is included in a multi-state model, prediction needs extra attention. Continuing with the example with cognitive function, say that the aim is to estimate residual life expectancies as discussed in Chapter 7. With misclassification present, life expectancies can be estimated conditional on observed state or conditional on latent state. When the objective is to make a statement with regard to the whole population, for example concerning the planning of future health care, estimating life expectancies conditional on latent state makes more sense as need for care will be induced by the true latent state. However, when state-specific life expectancies for a given individual are the primary quantities of interest, it seems more reasonable to base the estimation on current observed state and taking into account the potential misclassification (Van den Hout et al., 2014).

There is an extensive literature on methods for hidden Markov models. In addition to the references above, see, for example, Vermunt et al. (1999) and Bartolucci et al. (2012) for discrete-time Markov processes. Bartolucci et al. provide a literature review in which they also stress the link with hidden Markov models as used in the analysis of time-series.

For continuous-time Markov processes with extended models for the misclassification, see, for example, Jackson and Sharples (2002) and Van den Hout and Matthews (2008).

8.6 Smoothing splines and scoring

For multi-state survival data, Section 4.3 defines parametric time-dependent hazard regression models for transition intensities. As an alternative, this section discusses hazard regression models with semi-parametric time dependency.

Compared to semi- and non-parametric models, the two most obvious advantages of parametric models are *(i)* the limited number of parameters and *(ii)* the direct way in which the models can be used for prediction. There is of course also an inherent danger with parametric models: the chosen parametric shape may be too simple for the data at hand, in which case prediction is unreliable—especially when the prediction goes beyond the time range in the data.

There are many ways in which a semi-parametric model can be defined. Here we limited the discussion to B-splines which allow for flexible modelling of time dependency in a multi-state process.

When applied to time t as a variable, a B-spline is a piecewise polynomial function in t. The next illustration follows the general examples in Eilers and Marx (1996). A B-spline of degree 1 is defined using three knots $t_1 < t_2 < t_3$. The spline is zero on $t < t_1$ and on $t > t_3$. On $[t_1, t_3]$ it consists of two linear pieces: one from t_1 to t_2 and one from t_2 to t_3. The left-top graph in Figure 8.5 depicts this spline. A B-spline of degree 2 is defined using three quadratic pieces and four knots $t_1 < t_2 < t_3 < t_4$, and has a first derivative at any t; see top-left graph in Figure 8.5.

A basis of B-splines is often used for scatter-plot smoothing. Figure 8.5 illustrates this for toy data with a time dependency. In the left-bottom graph, the basis is defined by three B-splines of degree 1, where each spline is defined on the same range; in this case time interval $[0, 10]$. The spline which is depicted with a positive value at $t = 0$ is zero on $[t_2, t_3]$; the spline with a positive value at $t = 10$ is zero on $[t_1, t_2]$. The fit to the toy data is derived by using three parameters, one for each spline. The parameters scale the splines and the resulting curve is the sum of the three scaled splines. The individual scaling of the three splines is not depicted in Figure 8.5. When B-splines of degree 2 are used, the fitted curve is smooth in the sense that it has a first derivative. For the same toy data, this is illustrated in the right-bottom graph in Figure 8.5, where the basis is defined by five splines of degree 2.

A basis of B-splines can also be used to model the effect of a covariate in a regression model. Applications with continuous-time multi-state models can be found in Kneib and Hennerfeind (2008) and Titman (2011).

A B-spline expression of the baseline hazard for transition $r \to s$ is

$$q_{rs.0}(t) = \exp\left(\sum_{\ell=1}^{L} \alpha_{rs.\ell}\mathbf{B}_\ell(t)\right), \tag{8.13}$$

where—in this case—the choice of the number of splines L is the same for all

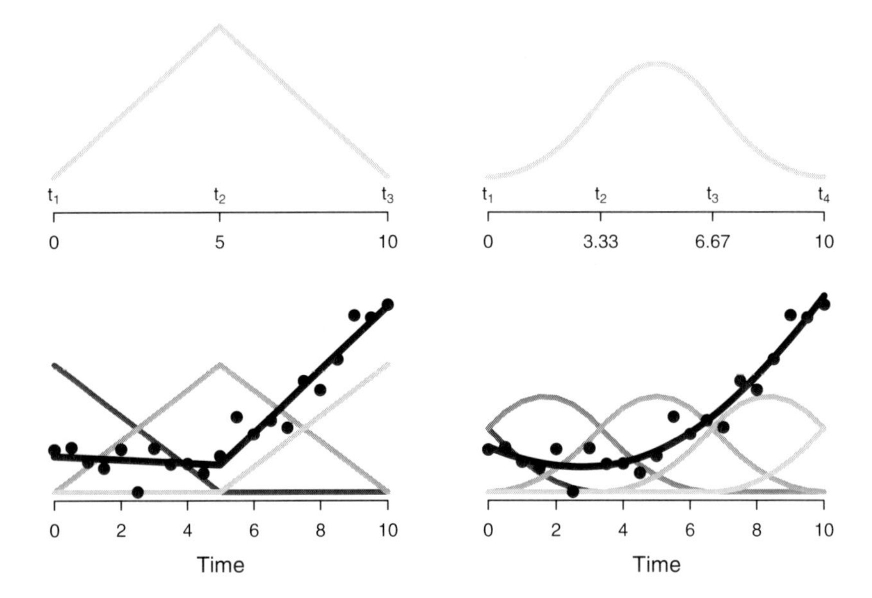

Figure 8.5 *Examples of B-splines. Top left: a spline of degree 1; top right: a spline of degree 2. For toy data, bottom left: fit using three splines of degree 1; bottom right: fit using five splines of degree 2.*

transitions $r \rightarrow s$, but the α-parameters are not. Expression (8.13) illustrates that the α-parameters scale the splines and that the resulting curve for the intensities is the sum of the scaled splines on the log-linear scale. It is also possible to define the spline basis directly on the scale of the intensities, but this implies restrictions on the α-parameters to enforce that the intensities are positive (Titman, 2011).

As with the parametric models for time-dependent intensities, hybrid models can be defined using the above. An example for a multi-state survival model is using Gompertz distribution for transitions into the dead state, and B-splines for transitions between living states.

Even if the interest lies in long-term prediction with a time span beyond the range of the data, splines can still be useful. An example of long-term prediction is the estimation of residual life expectancies as discussed in Chapter 7. Typically, in a longitudinal survey, data on centenarians—if available at all—are scarce. In such a situation, fitting splines for age range beyond 100 is a problem and one has to choose a parametric hazard model. It is also possible that there is enough information throughout the age range in the data, but that this age range does not include the assumed maximum age in the population. Also in this case, if the aim is to estimate life expectancies a parametric hazard model is needed. However, it may still be of interest to compare a fitted parametric distribution for the hazard with a semi-parametric shape fitted using a spline basis. If the semi-parametric shape is close to the parametric shape, then this can be seen as a validation of the distributional assumption which underlies the prediction based on the parametric model.

Maximum likelihood estimation of multi-state survival models which include B-spline formulations such as (8.13) can be undertaken by applying the scoring algorithm as specified in Section 4.6 when the time dependency is approximated piecewise-constantly. In this case, the only aspect of the algorithm that changes due to the use of splines is the first-order derivative of the generator matrix $\mathbf{Q}(t)$ with respect to the model parameters. This derivative, however, is straightforward for the B-splines in (8.13) since the spline expressions are linear in the α-parameters.

8.6.1 Example: ELSA study revisited

We illustrate the use of semi-parametric models for checking functional form as specified by parametric shapes.

The English Longitudinal Study of Ageing (ELSA) is are introduced in Section 4.8.1. For $N = 1000$ individuals in the age range 50 up to 90 years old, longitudinal outcomes are available on the number of words remembered in a delayed recall from a list of 10. For this proxy for cognitive function, and for death, a five-state survival model is specified; see Figure 4.2. The final model in the data analysis in Section 4.8.2 is Model I, which is given by

$$\begin{aligned}
q_{rs}(t) &= \exp\left(\beta_{rs.0} + \xi_{rs}t + \beta_{rs.1}sex + \beta_{rs.2}education\right) \\
&\qquad\qquad \text{for } (r,s) \in \{(1,2),(2,3),(3,4)\} \\
q_{rs}(t) &= \exp\left(\beta_{rs.0}\right) \qquad \text{for } (r,s) \in \{(2,1),(3,2),(4,3)\} \\
q_{rs}(t) &= \exp\left(\beta_{rs.0} + \xi_{D}t + \beta_{D.1}sex\right) \\
&\qquad\qquad \text{for } (r,s) \in \{(1,5),(2,5),(3,5),(4,5)\},
\end{aligned}$$

where *sex* is 0/1 for women/men, and *education* is 0/1 for the lower/higher level of education. This model has AIC = 7703.4.

In a study of change of cognitive function, interest is often focussed on the onset of cognitive impairment. Given that state 4 is defined for not being able to remember more than one word, we consider here the transition from state 3 to state 4 and investigate the B-spline alternative for the Gompertz hazard. For transition $3 \to 4$ only, we replace the hazard model by

$$q_{34.0}(t) = \exp\left(\sum_{\ell=1}^{L} \alpha_{34.\ell}\mathbf{B}_\ell(t) + \beta_{34.1}\,sex + \beta_{34.2}\,education\right).$$

For $L = 4$, this model has 23 parameters and AIC = 7702.1, which is lower than for Model 1 but the difference is rather minimal. There is no gain in using more splines; for $L = 5$, AIC = 7703.9, and for $L = 6$, AIC = 7704.9.

Figure 8.6 shows the fitted time dependency of the transition intensities for women with a higher level of education. As expected, there is an increase of risk of progression over the years to a lower cognitive function. The parametric shape is close to the time dependency estimated using B-splines. The discrepancy at the edges of the curves should be interpreted with care, since the estimation with B-splines is not optimal at the boundaries of the age range. Overall the comparison provides some evidence that the Gompertz is a good choice for the transition $3 \to 4$.

8.6.2 *More on the use of splines*

B-splines are a popular tool for smoothing, and software to define the splines is directly available. However, there are good alternatives. For multi-state models for interval-censored data, Joly et al. (1998) use twice differentiable M-splines, which are a variant of B-splines, and I-splines; see also Commenges et al. (1998) and Commenges (2002). For survival analysis, Aalen et al. (2008, Section 4.2.1) use the kernel method for a hazard model and discuss an example with the Epanechnikov kernel.

As long as the chosen spline basis for time t induces a straightforward first derivative of the generator matrix $\mathbf{Q}(t)$, the scoring algorithm can be used.

With any spline basis, the first problem is how many splines to use. One way to circumvent this problem is to use many and include a penalty for over-smoothing in the likelihood. One reason to choose B-splines is that a lot of work on smoothing penalties is available. For generalised linear models this is discussed by Eilers and Marx (1996). For multi-state models, penalised B-splines are used by Kneib and Hennerfeind (2008). Work on penalties is also available for other splines; see, for example, Joly et al. (1998) for multi-state models with M-splines and I-splines.

8.7 Semi-Markov models

The models so far in this book are all based on a conditional Markov assumption. Recall that a multi-state process is called a *Markov chain* if all

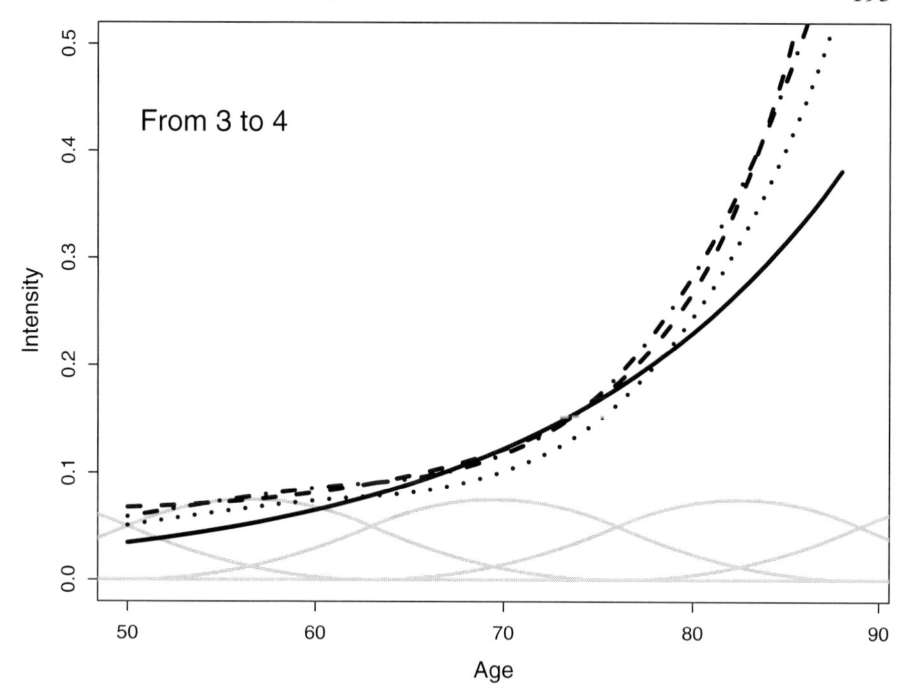

Figure 8.6 *For transition from state 3 to 4 in the ELSA data, estimated intensities for women with a higher level of education. Solid line for parametric Gompertz hazard, long-dashed line for B-spline model with $L = 4$, short-dashed line for $L = 5$, and dot-dashed line for $L = 6$. In grey the (unscaled) B-spline basis for $L = 5$.*

information about the future is contained in the present state. As stated already in Section 1.1, most of the multi-state modelling in this book is not Markovian in the strict sense; by linking time and values of covariates to the risk of a transition, information about the future is contained not only in the present state, but also in current time and additional background characteristics. More formally, the models assume that the process after time t is Markovian conditional on t and covariate values at t. Commenges (1999) calls this the *partial Markov condition*; the history of the process can be summarised by observations at t.

The conditional Markov assumption is an important aspect of the statistical modelling in this book as it allows the construction of the likelihood function as specified in Section 4.5. Although the data analyses in the examples, and many other analyses in the literature, show that statistical models based upon a Markov assumption can provide good and useful descriptions of multi-state processes, it is still of interest and importance to consider models that go beyond the Markov assumption.

For a discrete-time process $\{Y_n|\ n = 0, 1, 2, \ldots\}$, a Markov chain of order ν is a process such that

$$P(Y_J = y_J | Y_{J-1} = y_{J-1}, \ldots, Y_1 = y_1)$$
$$= P(Y_J = y_J | Y_{J-1} = y_{n-1}, \ldots, Y_{J-\nu} = y_{J-\nu}) \quad \text{for} \quad J > \nu.$$

In words, the future state depends on the past ν states. For the discrete-time processes in Section 8.1, the Markov chains are assumed to be of first order.

For a continuous-time process $\{Y_t | t \in (0, \infty)\}$, the definition of the order of the Markov chain is not applicable since time is not measured in steps. In this case, the Markov assumption at time t implies that the future of the process is conditional on the present state Y_t only.

Especially in biostatistics, it is easy to envisage a disease process in continuous time that is not Markovian. Consider, for example, the progressive three-state illness-death model. If time spent in the ill-health state affects the risk of a transition to the dead state, then the process is not Markovian.

When multi-state data are interval censored, models for non-Markovian continuous-time processes can be quite complex. Continuing with the progressive model example above, if the time of transition from the healthy state to the ill-health state is not observed, then time spent in the ill-health state is also not observed. If the three-state illness-death process is reversible in the sense that transitions are possible from the ill-health state back to the healthy state, the interval censoring is even more problematic as more than one transition can take place between two observation times.

A multi-state process for a state is called *semi-Markov* when the risk of leaving the state depends on the time spent in the state but not on the history of the process before entering the state.

More formally for a continuous-time process, let the jump times for a series $Y_1, Y_2, ..Y_J$ of distinct states be given by $S_1 = 0, S_2, ..., S_J$, respectively. So, for example, the process is in state Y_3 during time interval $[S_3, S_4)$. The stochastic process $\{Y_t | t \in (0, \infty)\}$ is a semi-Markov process if it has the Markov property at each of the jump times $S_1, S_2, ..., S_J$; see, for example, Kulkarni (2011, Section 5.4).

There is a long history of fitting semi-Markov models in biostatistics. Examples are Weiss and Zelen (1965), who discuss multi-state survival models in clinical trials, and Lagakos et al. (1978), who discusses how to deal with right censoring. Semi-Markov models are also discussed in Cox and Miller (1965, Section 9.3) as special cases of renewal models. Extensions and more recent applications can be found in Hougaard (2000), Foucher et al. (2010), Titman and Sharples (2010b), Kapetanakis et al. (2013), and Gillaizeau et al. (2015).

This section discusses an example of a semi-Markov model for a progressive three-state illness-death model in the presence of interval-censored transitions from state 1 to state 2. To limit the complexity of the model, we assume that all individuals in the data are in state 1 at baseline age, that the conditional Markov assumption holds for leaving state 1, and that the process is semi-Markov for state 2. The example serves to highlight the basic approach as discussed in, for example, Hougaard (2000, Chapter 6) and Foucher et al. (2010). The approach consists of integrating out the unobserved time of entering state 2 in the likelihood function contributions.

We choose Gompertz hazards and ignore covariates. The model for $t > 0$ is thus given by

$$\begin{aligned} q_{12}(t) &= \exp(\beta_{12} + \xi_{12}t) \\ q_{13}(t) &= \exp(\beta_{13} + \xi_{13}t) \\ q_{23}(t|u) &= \exp(\beta_{23} + \xi_{23}(t-u)), \end{aligned} \tag{8.14}$$

where u is the time of the transition into state 2. Note that the hazard for $2 \to 3$ conditional on u is again Gompertz with parameters $\beta_{23} + \xi_{23}u$ and ξ_{23}. By specifying ξ_{23} as the effect of $t-u$, parameter ξ_{23} is the effect of time spent in state 2 on the hazard of moving to state 3.

Because of the interval censoring, u is not observed. We start by distinguishing the four possible patterns of observed states within individual trajectories:

Pattern A: $\boxed{1} \cdots \boxed{1}$

Pattern B: $\boxed{1} \cdots \boxed{1}\boxed{2} \cdots \boxed{2}$

Pattern C: $\boxed{1} \cdots \boxed{1}\boxed{3}$

Pattern D: $\boxed{1} \cdots \boxed{1}\boxed{2} \cdots \boxed{2}\boxed{3}$

The dots in the above diagram mean that the number of observations of the same state may vary. For example, both

$\boxed{1}\boxed{1}\boxed{2}$ and $\boxed{1}\boxed{2}\boxed{2}$

are realisations of pattern B.

For each of the patterns, we give a data example with the corresponding likelihood contribution. Pattern A is straightforward because of the Markov assumption for state 1. The challenge is to deal with patterns B and D, where we know that there was a transition from state 1 to state 2, and pattern C, where there might have been a transition from state 1 to state 2 before death.

Say an individual with pattern A has observations at times t_1, t_2, and t_3. It follows that

$$P(Y_3 = 1, Y_2 = 1 | Y_1 = 1) = P(Y_3 = 1 | Y_2 = 1) P(Y_2 = 1 | Y_1 = 1).$$

Say an individual with pattern B has observed states 1, 2, and 2, at times t_1, t_2, and t_3, respectively. Integrating out the time of transition from state 1 to state 2 yields

$$\begin{aligned} P(Y_3 = 2, Y_2 = 2 | Y_1 = 1) &\\ &= \int_{t_2}^{t_3} P(Y_3 = 2, Y_2 = 2, u | Y_1 = 1) du \\ &= \int_{t_2}^{t_3} P(Y_3 = 2, Y_2 = 2 | u, Y_1 = 1) f(u | Y_1 = 1) du \\ &= \int_{t_2}^{t_3} P(Y_3 = 2, Y_2 = 2 | u) q_{12}(u) P(Y_u = 1 | Y_1 = 1) du \\ &= \int_{t_2}^{t_3} P(Y_3 = 2 | u) q_{12}(u) P(Y_u = 1 | Y_1 = 1) du. \end{aligned}$$

In this derivation, we use $f(t) = h(t)S(t)$ for hazard function h and survivor function S; see Section 2.2. Note that because the process is progressive, $P(Y_3 = 2, Y_2 = 2 | u) = P(Y_3 = 2 | u)$.

For pattern C, say an individual has observed states 1 and 3, at times t_1 and t_2, respectively. Let t_2^- denote the time just before death. The state at t_2^- is either state 1 or state 2. With the law of total probability it follows that

$$\begin{aligned} P(Y_2 = 3 | Y_1 = 1) &\\ &= P(Y_2 = 3, Y_{t_2^-} = 1 | Y_1 = 1) + P(Y_2 = 3, Y_{t_2^-} = 2 | Y_1 = 1) \\ &= P(Y_{t_2^-} = 1 | Y_1 = 1) q_{13}(t_2) \\ &\quad + \int_{t_2}^{t_3} q_{23}(t_2 | u) P(Y_{t_2^-} = 2 | u) q_{12}(u) P(Y_u = 1 | Y_1 = 1) du. \end{aligned}$$

For an individual with pattern D and observed states 1, 2, and 3, at times t_1, t_2 and t_3, respectively, we get

$$\begin{aligned} P(Y_3 = 3, Y_2 = 2 | Y_1 = 1) &\\ &= \int_{t_2}^{t_3} q_{23}(t_3 | u) P(Y_{t_3^-} = 2 | u) q_{12}(u) P(Y_u = 1 | Y_1 = 1) du. \end{aligned}$$

These four data examples cover the range of potential individual trajectories in the data since any other series of states is just a simple variation or extension of one of these examples.

If we use a piecewise-constant approximation defined by the data at hand (Section 4.4), then there are closed-form expressions for all the relevant probabilities. Without loss of generality, for states (Y_1, Y_2) at times (t_1, t_2), we get

$$\begin{aligned} P(Y_2 = 1|Y_1 = 1) &= \exp\big(-(q_{12}(t_1) + q_{13}(t_1))(t_2 - t_1)\big) \\ P(Y_2 = 2|u) &= \exp\big(-q_{23}(u|u)(t_2 - u)\big). \end{aligned}$$

Note that time u of the transition $1 \rightarrow 2$ is not given in the data, but is defined within the integral and can thus be used in the approximation of the integrand. If there are multiple observations of state 2 in an individual trajectory, then the time intervals in the data can be used to define the piecewise-constant approximation. For example if state 1 is observed at time t_1, and state 2 is observed at t_2 and t_3, then

$$\begin{aligned} P(Y_3 = 2|u) &= P(Y_3 = 2|Y_2 = 2, u)P(Y_2 = 2|u) \\ &= \exp\big(-q_{23}(t_2|u)(t_3 - t_2)\big)\exp\big(-q_{23}(u|u)(t_2 - u)\big). \end{aligned}$$

In case time intervals in the data are long relative to the volatility of the process, the piecewise-constant approximation can also be defined by using an imposed grid; see Section 4.4.

Likelihood inference can be undertaken by approximating the integrals using the trapezoidal rule or any other numerical method for integration, and by using a general-purpose optimiser to maximise the likelihood function.

There are no general guidelines for the performance of the piecewise-constant approximation or the number of nodes needed in the numerical integration. However, the simulation scheme in Section 3.6 can easily be adapted for semi-Markov models. Performance can thus be investigated using simulated data. For example, we ran a simulation study for the Gompertz model (8.14) specified by $(\beta_{12}, \beta_{13}, \beta_{23}) = (-5, -4, -3)$ and $(\xi_{12}, \xi_{13}, \xi_{23}) = (0.1, 0.15, 0.4)$ for time scale age in years. Good results were obtained (not reported here) for a sample size of 500 individuals (all with baseline state 1), 12 years of yearly observations, and 15 nodes in the trapezoidal rule.

Appendix A

Matrix $\mathbf{P}(t)$ When Matrix $\mathbf{Q}$ Is Constant

This appendix discusses the computation of transition probability matrix $\mathbf{P}(t)$, for time $t > 0$, conditional on a constant generator matrix $\mathbf{Q}$. The methods in this appendix are used throughout the book whenever a piecewise-constant approximation of time-dependent transition intensities is applied.

There are no new results in this appendix. The aim is to explain the methods, to illustrate the generality of the approach, and to provide an overview. The material in this appendix is a bit scattered over the literature and not always provided in detail; see, for instance, Norris (1997), Moler and Van Loan (2003), Welton and Ades (2005, and its supplement), Ross (2010), Kulkarni (2011), and Jackson (2011).

For a continuous-time Markov process with time-homogeneous $D \times D$ generator matrix $\mathbf{Q}$, the transition probability matrix $\mathbf{P}(t)$ for time $t > 0$ is the solution to the Kolmogorov backward equation $\mathbf{P}'(t) = \mathbf{QP}(t)$ subject to $\mathbf{P}(0) = \mathbf{I}_D$, where $\mathbf{I}_D$ is the identity matrix. The solution is the matrix exponential

$$\mathbf{P}(t) = e^{t\mathbf{Q}} = \sum_{k=1}^{\infty} \frac{(t\mathbf{Q})^k}{k!}. \tag{A.1}$$

If the $D \times D$ matrix $\mathbf{Q}$ has D linearly independent eigenvectors, then the matrix exponential can be expressed using the eigenvalue decomposition of $\mathbf{Q}$. An eigenvalue decomposition of $\mathbf{Q}$ implies that we can write $\mathbf{Q} = \mathbf{ABA}^{-1}$, where the columns of the $D \times D$ matrix $\mathbf{A}$ are D linearly independent eigenvectors of $\mathbf{Q}$, and $\mathbf{B}$ is a diagonal matrix with the D eigenvalues of $\mathbf{Q}$ as diagonal entries. If $\mathbf{Q}$ can be decomposed in this way, then $\mathbf{P}(t)$ is given by

$$\mathbf{P}(t) = \mathbf{A}\ \mathrm{diag}\left(e^{b_1 t}, ..., e^{b_D t}\right)\ \mathbf{A}^{-1},$$

where $b_1, ..., b_D$ are the eigenvalues of $\mathbf{Q}$, and $\mathrm{diag}(x_1, ..., x_D)$ denotes the $D \times D$ diagonal matrix with diagonal entries $x_1, ..., x_D$. Because of the decomposition of $\mathbf{Q}$, the matrix exponential for $\mathbf{P}(t)$ has thus been reduced to a series of scalar exponentials, which simplifies the computation of $\mathbf{P}(t)$ considerably.

Not all $D \times D$ matrices have D linearly independent eigenvectors. Consider the following generator matrix for a three-state progressive survival model

$$\mathbf{Q} = \begin{pmatrix} -(q_{12}+q_{13}) & q_{12} & q_{13} \\ 0 & -q_{23} & q_{23} \\ 0 & 0 & 0 \end{pmatrix} = \begin{pmatrix} -1 & 1/2 & 1/2 \\ 0 & -1 & 1 \\ 0 & 0 & 0 \end{pmatrix}.$$

This matrix has eigenvalues -1, -1, and 0, with eigenvectors $(1,0,0)^{\top}$, $(-1,0,0)^{\top}$, and $(\sqrt{1/3}, \sqrt{1/3}, \sqrt{1/3})$, respectively. For this specific $\mathbf{Q}$, the matrix exponential cannot be expressed using the eigenvalue decomposition.

On the other hand, not all $D \times D$ matrices that can be decomposed have D distinct eigenvalues. A simple example is the identity matrix, which has D linearly independent eigenvectors all with eigenvalue 1.

For some matrices that are restricted in size, eigenvalues can be computed symbolically, and a closed-form solution can be derived for $\mathbf{P}(t)$. For the closed-form solutions in this appendix, eigenvalues were derived using a website for symbolic matrix calculations: WIMS at

`http://wims.unice.fr/wims/en_tool~linear~matrix.html`

by Gang Xiao (Version 2.20, 1997-1999). For the applications in the book without a closed-form solution, the R function `eigen` was used. This function uses LAPACK routines; see

`http://www.netlib.org/lapack/lug/lapack_lug.html`

by Anderson et al. (1999).

As a general method to compute a matrix exponential, using eigenvectors does not come highly recommended; see Moler and Van Loan (2003). A square matrix is called *defective* if it does not have a complete basis of eigenvectors, and is therefore not diagonalisable. As stated by Moler and Van Loan (2003), difficulties occur in practice when a matrix is "nearly" defective. In terms of generator matrices, this means that taking the inverse of $\mathbf{A}$ to obtain $\mathbf{Q} = \mathbf{ABA}^{-1}$ is subject to numerical problems and the result may be inaccurate.

There are many alternatives to using eigenvectors. A number of these methods are implemented in the R package expm developed by Goulet et al. (2014). If a general-purpose optimiser is used to fit a multi-state survival model, then any of these alternatives can be used; see Section 4.5. The scoring algorithm in Section 4.6 is based upon using eigenvectors.

For a $D \times D$ generator matrix $\mathbf{Q}$, the absence of D linearly independent eigenvectors can be induced by severe model restrictions. The example above for $D = 3$ is relevant for the restriction $q_{12} = q_{13}$. In the practice of fitting multi-state survival models a restriction such as $q_{12} = q_{13}$ is not realistic in most cases where state 3 is the dead state. Furthermore, because hazard-specific regression models will often include several hazard-specific parameters, it is also not to be expected that entries in $\mathbf{Q}$ are equal when a model is fitted to data. The exception of this can occur when starting values are chosen in a specific, restrictive way. In the example for $D = 3$, say, the model is $q_{rs}(t) = \exp(\beta_{rs} + \xi_{rs}t)$ for $(r,s) \in \{(1,2),(1,3),(2,3)\}$. Given this Gompertz hazard model, any optimisation method using eigenvalue decomposition will run into problems when starting values are specified by $(\log(.5), \log(.5), \log(1))$ for $(\beta_{12}, \beta_{13}, \beta_{23})$ and by zeroes for the ξ-parameters.

A.1 Two-state models

The two-state survival model is the standard model discussed in Chapter 2:

[1] → [2]

$$\mathbf{Q} = \begin{pmatrix} -q_{12} & q_{12} \\ 0 & 0 \end{pmatrix}.$$

For the one event of interest, the constant hazard is q_{12}, for $q_{12} > 0$. This is the hazard of the exponential distribution in Section 2.3.1. The transition probability is $p_{12}(t) = F(t)$, where F is the cumulative distribution function of the exponential distribution. Hence $p_{12}(t) = 1 - \exp(-q_{12}t)$, and

$$\mathbf{P}(t) = \begin{pmatrix} e^{-q_{12}t} & 1 - e^{-q_{12}t} \\ 0 & 1 \end{pmatrix}.$$

This matrix can also be derived by solving $\mathbf{P}'(t) = \mathbf{QP}(t)$ subject to $\mathbf{P}(0) = \mathbf{I}_2$.

The two-state model with two transitions is not featured in the book, but will serve here as an introduction to more complex models:

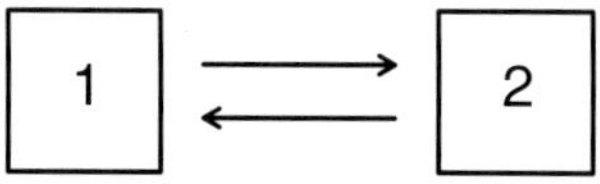

[1] ⇄ [2]

$$\mathbf{Q} = \begin{pmatrix} -q_{12} & q_{12} \\ q_{21} & -q_{21} \end{pmatrix}.$$

It follows that the Kolmogorov backward equation is the differential equation $p'_{11}(t) = q_{21} - (q_{21} + q_{12})p_{11}(t)$. Tackling this equation head-on, a solution can be found for $p_{11}(t)$. Using this solution, probability $p_{22}(t)$ can be derived. These two probabilities define the 2×2 transition matrix $\mathbf{P}(t)$ completely. It follows that

$$\mathbf{P}(t) = \begin{pmatrix} \frac{q_{21}+q_{12}e^{-(q_{21}+q_{12})t}}{q_{21}+q_{12}} & 1-p_{11}(t) \\ 1-p_{22}(t) & \frac{q_{12}+q_{21}e^{-(q_{21}+q_{12})t}}{q_{21}+q_{12}} \end{pmatrix}.$$

An alternative derivation is to use $\mathbf{P}(t) = \exp(t\mathbf{Q})$. First diagonalise $\mathbf{Q}$ as

$$\mathbf{Q} = \mathbf{U} \begin{pmatrix} 0 & 0 \\ 0 & -(q_{21}+q_{12}) \end{pmatrix} \mathbf{U}^{-1}, \quad \text{for} \quad \mathbf{U} = \begin{pmatrix} 1 & 1 \\ 1 & \frac{-q_{21}}{q_{12}} \end{pmatrix},$$

where diagonal entries 0 and $-(q_{21}+q_{12})$ at the left-hand side are the eigenvalues of $\mathbf{Q}$, and the columns of $\mathbf{U}$ are the corresponding eigenvectors. The matrix exponential is now given by

$$\begin{aligned} \mathbf{P}(t) = \sum_{k=1}^{\infty} \frac{(t\mathbf{Q})^k}{k!} &= \mathbf{U} \left[\sum_{k=1}^{\infty} \frac{1}{k!} \begin{pmatrix} 0^k & 0 \\ 0 & (-(q_{21}+q_{12})t)^k \end{pmatrix} \right] \mathbf{U}^{-1} \\ &= \mathbf{U} \begin{pmatrix} 1 & 0 \\ 0 & e^{-(q_{21}+q_{12})t} \end{pmatrix} \mathbf{U}^{-1}. \end{aligned}$$

A.2 Three-state models

The three-state survival progressive model is discussed in Chapter 2:

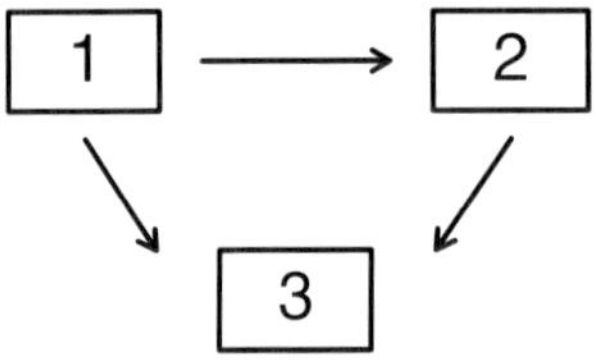

$$\mathbf{Q} = \begin{pmatrix} -(q_{12}+q_{13}) & q_{12} & q_{13} \\ 0 & -q_{23} & q_{23} \\ 0 & 0 & 0 \end{pmatrix}.$$

A derivation of the transition probabilities is given in Section 3.2.1. An alternative derivation is to use the eigenvalue decomposition of $\mathbf{Q}$ given by

$$\mathbf{Q} = \mathbf{U} \begin{pmatrix} -q_{23} & 0 & 0 \\ 0 & -(q_{13}+q_{12}) & 0 \\ 0 & 0 & 0 \end{pmatrix} \mathbf{U}^{-1}, \quad \text{for} \quad \mathbf{U} = \begin{pmatrix} 1 & 1 & 1 \\ \frac{q_{13}+q_{12}-q_{23}}{q_{12}} & 0 & 1 \\ 0 & 0 & 1 \end{pmatrix}.$$

Matrix $\mathbf{P}(t) = \exp(t\mathbf{Q})$ can now be derived by matrix multiplication; see the two-state case above. It follows that

$$\mathbf{P}(t) = \begin{pmatrix} e^{-(q_{12}+q_{13})t} & \frac{q_{12}\left(e^{-q_{23}t} - e^{-(q_{12}+q_{13})t}\right)}{q_{12}+q_{13}-q_{23}} & 1-p_{11}(t)-p_{12}(t) \\ 0 & p_{22}(t) & 1-e^{-q_{23}t} \\ 0 & 0 & 1 \end{pmatrix}$$

for $q_{23} \neq q_{12}+q_{13}$. If $q_{23}=q_{12}+q_{13}$, then a solution can be derived by using the Kolmogorov backward equation $\mathbf{P}'(t) = \mathbf{QP}(t)$, where $\mathbf{Q}$ is completely defined by q_{12} and q_{13}. The differential equation is solved by first deriving $p_{11} = p_{22} = \exp(-q_{23}t)$ and then solving $p'_{12} = -q_{23}p_{12} + q_{12}p_{22}$. This leads to $p_{12}(t) = q_{12}t\exp(-q_{23}t)$.

A progressive three-state model with two absorbing states is called a competing risks model. For this model, which is not discussed explicitly in the book, we have

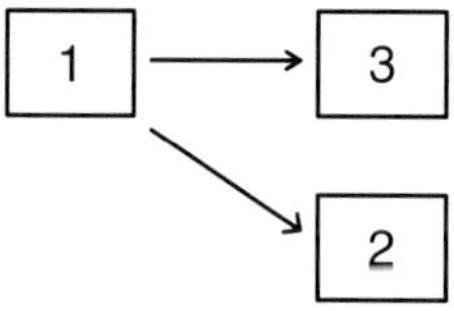

1 → 3, 1 → 2

$$\mathbf{Q} = \begin{pmatrix} -(q_{12}+q_{13}) & q_{12} & q_{13} \\ 0 & 0 & 0 \\ 0 & 0 & 0 \end{pmatrix}.$$

The eigenvalue decomposition is given by

$$\mathbf{Q} = \mathbf{U}\begin{pmatrix} -(q_{12}+q_{13}) & 0 & 0 \\ 0 & 0 & 0 \\ 0 & 0 & 0 \end{pmatrix}\mathbf{U}^{-1} \quad \text{with} \quad \mathbf{U} = \begin{pmatrix} 1 & 1 & 0 \\ 0 & 0 & 1 \\ 0 & \frac{q_{12}+q_{13}}{q_{13}} & -\frac{q_{12}}{q_{13}} \end{pmatrix}.$$

This leads to

$$\mathbf{P}(t) = \begin{pmatrix} e^{-(q_{12}+q_{13})t} & \frac{q_{12}}{q_{12}+q_{13}}\left(1-e^{-(q_{12}+q_{13})t}\right) & \frac{q_{13}}{q_{12}+q_{13}}\left(1-e^{-(q_{12}+q_{13})t}\right) \\ 0 & 1 & 0 \\ 0 & 0 & 1 \end{pmatrix}.$$

This is an example of a $\mathbf{Q}$ matrix that is decomposable even though the eigenvalues are not distinct. Using the eigenvalue decomposition in this case is for illustration only. As an easier alternative, note that the time to leave state 1 is exponentially distributed with rate $q_{12}+q_{13}$, so $p_{11}(t)$ is the survivor function at time t. Given a transition out of state 1, the state is going to be state 2 or state 3, with the overall probability of moving weighted by $q_{12}/(q_{12}+q_{13})$ and $q_{13}/(q_{12}+q_{13})$, respectively.

The three-state survival model with recovery is a specialisation of the general model discussed in Chapter 4:

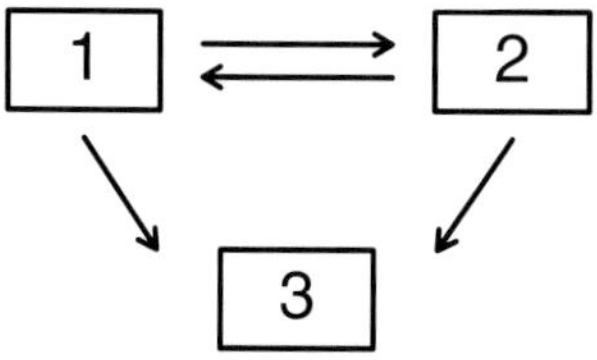

1 ⇄ 2, 1 → 3, 2 → 3

$$\mathbf{Q} = \begin{pmatrix} -(q_{12}+q_{13}) & q_{12} & q_{13} \\ q_{21} & -(q_{21}+q_{23}) & q_{23} \\ 0 & 0 & 0 \end{pmatrix}.$$

The following derivation of $\mathbf{P}(t)$ is based upon the supplement to Welton and Ades (2005). To keep the expression in the symbolic computation limited, it helps to define $\lambda_1 = q_{12} + q_{13}$ and $\lambda_2 = q_{21} + q_{23}$. The eigenvalue decomposition is then given by

$$\mathbf{Q} = \mathbf{U} \begin{pmatrix} \frac{-h-\lambda_1-\lambda_2}{2} & 0 & 0 \\ 0 & \frac{h-\lambda_1-\lambda_2}{2} & 0 \\ 0 & 0 & 0 \end{pmatrix} \mathbf{U}^{-1},$$

for

$$\mathbf{U} = \begin{pmatrix} 1 & 1 & 1 \\ \frac{-h+\lambda_1-\lambda_2}{2q_{12}} & \frac{h+\lambda_1-\lambda_2}{2q_{12}} & 1 \\ 0 & 0 & 1 \end{pmatrix},$$

where $h = \sqrt{4q_{12}q_{21} + (\lambda_1 - \lambda_2)^2}$. After computing the inverse of **U** symbolically, entries of the matrix exponential can be derived. We obtain

$$\begin{aligned}
p_{11}(t) &= \frac{(-\lambda_1 + \lambda_2 + h)e^{-\frac{1}{2}(\lambda_1+\lambda_2-h)t} + (\lambda_1 - \lambda_2 + h)e^{-\frac{1}{2}(\lambda_1+\lambda_2+h)t}}{2h} \\
p_{12}(t) &= \frac{q_{12}}{h}\left(e^{-\frac{1}{2}(\lambda_1+\lambda_2-h)t} - e^{-\frac{1}{2}(\lambda_1+\lambda_2+h)t}\right) \\
p_{13}(t) &= 1 - p_{11}(t) - p_{12}(t) \\
p_{12}(t) &= \frac{q_{21}}{h}\left(e^{-\frac{1}{2}(\lambda_1+\lambda_2-h)t} - e^{-\frac{1}{2}(\lambda_1+\lambda_2+h)t}\right) \\
p_{22}(t) &= \frac{(-\lambda_1 + \lambda_2 + h)e^{-\frac{1}{2}(\lambda_1+\lambda_2+h)t} + (\lambda_1 - \lambda_2 + h)e^{-\frac{1}{2}(\lambda_1+\lambda_2-h)t}}{2h} \\
p_{23}(t) &= 1 - p_{21}(t) - p_{22}(t) \\
p_{31}(t) &= 0, \qquad p_{32}(t) = 0, \qquad p_{33}(t) = 1,
\end{aligned}$$

where the expression for $p_{12}(t)$ is a simplified version of the expression in the supplement to Welton and Ades (2005).

The three-state without death is discussed in Section 8.2.1:

1 ⇄ 2 ↑↓ 3

$$\mathbf{Q} = \begin{pmatrix} -q_{12} & q_{12} & 0 \\ q_{21} & -(q_{21} + q_{23}) & q_{23} \\ 0 & q_{32} & -q_{32} \end{pmatrix}.$$

For this model, deriving the eigenvalues and eigenvectors is possible symbolically, but deriving the corresponding closed-form expression for $\mathbf{P}(t)$ is not very inviting given the complexity of the expressions for the eigenvectors. Instead we illustrate two methods for the numerical derivation of $\mathbf{P}(t)$ for a given $\mathbf{Q}$. Consider the following two generator matrices:

$$\mathbf{Q}_1 = \begin{pmatrix} -0.10 & 0.10 & 0 \\ 0.05 & -0.10 & 0.05 \\ 0 & 0.02 & -0.02 \end{pmatrix} \quad \mathbf{Q}_2 = \begin{pmatrix} -2 & 2 & 0 \\ 0.05 & -0.10 & 0.05 \\ 0 & 2 & -2 \end{pmatrix}.$$

Let K be the finite number of summations to approximate the infinite summation (A.1) for the matrix exponential. Say that two numerical transition matrices are the same if the first six decimal places are the same for all entries to be compared.

For $\mathbf{Q}_1$ and $t = 1$, using the summation approximation with $K = 5$ produces the same $\mathbf{P}(t)$ as using the default setting in the function `eigen` for eigenvalue decomposition in R. For $t = 5$, the summation needs $K = 6$ to produce the same matrix as the decomposition method.

The matrix $\mathbf{Q}_2$ describes a process where duration of stay in states 1 and 3 is very short compared to duration of stay in state 2. For this matrix and $t = 1$, the summation needs $K = 13$ to produce the same matrix as the eigenvalue decomposition. The number of iterations needed in the summation increases rapidly with increasing t. For $t = 5$, the summation needs $K = 40$ to produce the same matrix as the decomposition method. For $t = 10$, $K = 70$ is needed. There are methods to automate the choice of K; see, for example, the uniformisation algorithm proposed by Kulkarni (2011) which is illustrated in Section 4.2.

The choice of the three-state models above is mainly driven by the applications in this book. There are additional three-state models; for example, the one with transitions $1 \rightarrow 3$ and $2 \rightarrow 3$ only. These models, however, can be dealt with along the same lines as set out above.

A.3 Models with more than three states

The four-state progressive survival model is discussed in Section 1.3.2. The diagram is

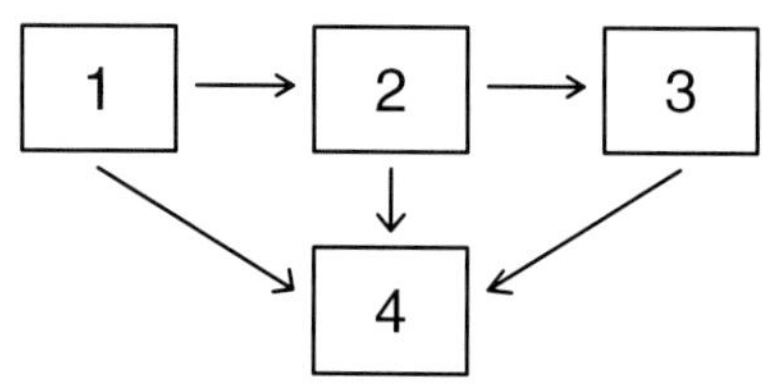

and the generator matrix is given by

$$\mathbf{Q} = \begin{pmatrix} -(q_{12}+q_{13}+q_{14}) & q_{12} & q_{13} & q_{14} \\ 0 & -(q_{23}+q_{24}) & q_{23} & q_{24} \\ 0 & 0 & -q_{34} & q_{34} \\ 0 & 0 & 0 & 0 \end{pmatrix}.$$

The eigenvalues of $\mathbf{Q}$ are $-q_{34}$, $-(q_{24}+q_{23})$, $-(q_{14}+q_{13}+q_{12})$, and 0. The corresponding eigenvectors are given by

$$\begin{pmatrix} 1 \\ \frac{q_{23}q_{34}-(q_{14}+q_{13}+q_{12})q_{23}}{h} \\ \frac{q_{34}^2-(q_{24}+q_{23}+q_{14}+q_{13}+q_{12})q_{34}+(q_{14}+q_{13}+q_{12})q_{24}+(q_{14}+q_{13}+q_{12})q_{23}}{-h} \\ 0 \end{pmatrix},$$

where $h = q_{13}q_{34} - q_{13}q_{24} - (q_{13}+q_{12})q_{23}$, and

$$\begin{pmatrix} 1 \\ \frac{-(q_{24}+q_{23}-q_{14}-q_{13}-q_{12})}{q_{12}} \\ 0 \\ 0 \end{pmatrix} \begin{pmatrix} 1 \\ 0 \\ 0 \\ 0 \end{pmatrix} \begin{pmatrix} 1 \\ 1 \\ 1 \\ 1 \end{pmatrix},$$

respectively.

It is clear that expressions quickly become complex once more than three states are considered. For the five-state progressive survival model (as the four-state model above but with one extra living state), the eigenvectors can be computed symbolically, but the expressions are awkwardly long.

For models with more than three states, it is up to the user to decide whether it is worthwhile to program closed-form expressions (which will be fast in computing time once coded) or to use numerical approximation instead (which will take longer in computing time, but can be carried out with available software).

Appendix B

Scoring for the Progressive Three-State Model

This appendix provides details for the scoring algorithm for the progressive three-state survival model when using piecewise-constant intensities. This adaptation of the algorithm in Kalbfleisch and Lawless (1985) is discussed in general in Section 4.6.

The scoring algorithm is based on the eigenvalue decomposition of the time-homogeneous 3×3 generator matrix $\mathbf{Q}$. The decomposition can be used to derive a closed-form expression for the transition matrix $\mathbf{P}(t)$ for given time t; see Appendix A.

Let $\boldsymbol{\theta}$ be the vector with the model parameters, and let θ_k denote entry k. The derivative of $\mathbf{P}(t)$ can be obtained as

$$\frac{\partial}{\partial \theta_k} \mathbf{P}(t) = \mathbf{A} \mathbf{V}_k \mathbf{A}^{-1},$$

where $\mathbf{A}$ is the matrix with the three distinct eigenvectors as columns, and $\mathbf{V}_k$ is the 3×3 matrix with (l,m) entry

$$\begin{cases} g_{lm}^{(k)} \big(\exp(b_l t) - \exp(b_m t)\big) / (b_l - b_m) & l \neq m \\ \\ g_{ll}^{(k)} t \exp(b_l t) & l = m, \end{cases}$$

where $g_{lm}^{(k)}$ is the (l,m) entry in $\mathbf{G}^{(k)} = \mathbf{A} \partial \mathbf{Q} / \partial \theta_k \mathbf{A}^{-1}$, and b_1, b_2, b_3 are the eigenvalues. The upshot of the above is that $\partial \mathbf{Q} / \partial \theta_k$ is in most cases easy to derive.

This general setup will be illustrated for the Gompertz hazard model $q_{rs}(t) = \exp(\beta_{rs} + \xi t)$, with $\boldsymbol{\theta} = (\beta_{12}, \beta_{13}, \beta_{23}, \xi)$. For this model, it follows that

$$\frac{\partial}{\partial \theta_1} \mathbf{Q}(t) = \frac{\partial}{\partial \beta_{12}} \mathbf{Q}(t) = \begin{pmatrix} -q_{12}(t) & q_{12}(t) & 0 \\ 0 & 0 & 0 \\ 0 & 0 & 0 \end{pmatrix},$$

$$\frac{\partial}{\partial \theta_2}\mathbf{Q}(t) = \frac{\partial}{\partial \beta_{13}}\mathbf{Q}(t) = \begin{pmatrix} -q_{13}(t) & 0 & q_{13}(t) \\ 0 & 0 & 0 \\ 0 & 0 & 0 \end{pmatrix},$$

$$\frac{\partial}{\partial \theta_3}\mathbf{Q}(t) = \frac{\partial}{\partial \beta_{23}}\mathbf{Q}(t) = \begin{pmatrix} 0 & 0 & 0 \\ 0 & -q_{13}(t) & q_{23}(t) \\ 0 & 0 & 0 \end{pmatrix},$$

and

$$\frac{\partial}{\partial \theta_4}\mathbf{Q}(t) = \frac{\partial}{\partial \xi}\mathbf{Q}(t) = \begin{pmatrix} -t\big(q_{12}(t)+q_{13}(t)\big) & tq_{12}(t) & tq_{13}(t) \\ 0 & -tq_{23}(t) & tq_{23}(t) \\ 0 & 0 & 0 \end{pmatrix}.$$

To specify the scoring algorithm, the notation in Section 3.5 is extended as follows. For individual i, the observed trajectory is the series of states $y_{i1},\ldots,y_{iJ_i}$ corresponding to times $t_{i1},\ldots,t_{iJ_i}$. The log-likelihood function conditional on the first state is thus given by

$$\log L = \sum_{i=1}^{N}\sum_{j=2}^{J_i}\log L_{ij} = \sum_{i=1}^{N}\left(\sum_{j=2}^{J_i-1}\log\left(p_{y_{ij-1}y_{ij}}(w_{ij})\right)\right) + \log\left(C(y_{J_i}|y_{J_i-1})\right),$$

where $w_{ij} = t_{ij} - t_{ij-1}$. Using the piecewise-constant approximation implies that the intensities do not change in $(t_{ij-1}, t_{ij}]$. Let $\mathbf{Q}_{ij-1}$ denote the intensity matrix for $(t_{ij-1}, t_{ij}]$. The first-order derivative (the *score*) is

$$S_k(\boldsymbol{\theta}) = \frac{\partial}{\partial \theta_k}\log L =$$

$$\sum_{i=1}^{N}\left(\sum_{j=2}^{J_i-1}\frac{1}{p_{y_{ij-1}y_{ij}}(w_{ij})}\frac{\partial}{\partial \theta_k}p_{y_{ij-1}y_{ij}}(w_{ij})\right) + \frac{1}{C(y_{J_i}|y_{J_i-1})}\frac{\partial}{\partial \theta_k}C(y_{J_i}|y_{J_i-1}).$$

It follows that for right censoring,

$$\frac{\partial}{\partial \theta_k}C(y_{J_i}|y_{J_i-1}) = \frac{\partial}{\partial \theta_k}p_{y_{iJ_i-1}1}(w_{ij}) + \frac{\partial}{\partial \theta_k}p_{y_{iJ_i-1}2}(w_{ij}),$$

and for known death times

$$\frac{\partial}{\partial \theta_k}C(y_{J_i}|y_{J_i-1}) = \frac{\partial}{\partial \theta_k}\left(p_{y_{iJ_i-1}1}(w_{ij})q_{13}\right) + \frac{\partial}{\partial \theta_k}\left(p_{y_{iJ_i-1}2}(w_{ij})q_{23}\right),$$

where q_{13} and q_{23} are entries of $\mathbf{Q}_{ij-1}$.

The score can now be obtained using the eigenvalue decomposition of

$\mathbf{Q}_{ij-1}$ to obtain $\frac{\partial}{\partial \theta_k}\mathbf{P}(t)$ as described above for a constant $\mathbf{Q}$. And the scoring algorithm can be implemented in a standard way; see Section 4.6.

To illustrate the speed of the algorithm, the Gompertz baseline hazards model (3.3) and the Weibull baseline hazards model (3.4) will be estimated using the piecewise-constant approximation. The data are from the Parkinson's disease study with $N = 233$ in Section 3.7, and the grid for piecewise-constant approximation is defined by the individual-specific time intervals as observed in the data.

To be able to maximise the log-likelihood over an unrestricted parameter space, the Gompertz model is reparametrised for $\lambda_{rs} > 0$ by

$$\begin{aligned} q_{rs}(t) &= \lambda_{rs}\exp(\xi_{rs}t) \\ &= \exp(\beta_{rs} + \xi_{rs}t) \quad \text{for} \quad (r,s) \in \{(1,2),(1,3),(2,3)\}, \end{aligned}$$

and the Weibull model for $\tau_{rs} > 0$ by

$$\begin{aligned} q_{rs}(t) &= \exp(\beta_{rs})\tau_{rs}t^{(\tau_{rs}-1)} \\ &= \exp(\beta_{rs} + \gamma_{rs})t^{(\exp(\gamma_{rs})-1)} \quad \text{for} \quad (r,s) \in \{(1,2),(1,3),(2,3)\}. \end{aligned}$$

After finding the maximum likelihood estimate for the unrestricted parameters, the delta method can be used to derive inference for λ_{rs} and τ_{rs}. Performance of the scoring algorithm will be compared to using the general-purpose optimiser `optim` in R.

It is possible to provide `optim` with the first derivative in the function call. That option will not be used here. Thus the scoring algorithm has an advantage as it makes use of the derivative in the optimisation. The comparison is not meant to criticise `optim`, which is a powerful tool in many situations.

For the scoring, convergence at iteration r is defined as

$$\sum_k |\theta_k^{(r-1)} - \theta_k^{(r)}| < \varepsilon,$$

for $\varepsilon = 1 \times 10^{-6}$. Both the Gompertz model and the Weibull model are formulated with six parameters so the summation in the convergence criterion is over $k = 1,..,6$.

Starting values for the Gompertz model are $\beta_{rs} = -5$ and $\xi_{rs} = 0$ for the relevant combinations of r and s. Convergence is achieved in 23 iterations. Using ε as the absolute convergence tolerance in optim, 53 iterations are needed when using the method by Broyden, Fletcher, Goldfarb, and Shanno (BFGS). In addition to these iterations, `optim` runs extra iterations to estimate the Hessian matrix. Using the scoring algorithm in this setting is about three times faster.

Starting values for the Weibull model are $\beta_{rs} = -9$ and $\gamma_{rs} = 1$. Convergence is achieved in 18 iterations using ε as given above. Maximisation using `optim` requires 97 iterations to obtain the optimum and a total time that is almost eight times slower than needed for the scoring.

Appendix C

Some Code for the R and BUGS Software

The software R is a free environment for statistical computing and graphics (R Core Team, 2013). All the graphics in this book and almost all the analyses are produced using R. The illustrative R code in this appendix is not optimised with respect to speed but is structured in such a way that it resembles the formulas in the book and can be understood even if knowledge of R is minimal. This appendix only provides the basic elements of the coding; for the full code please see the website which accompanies this book.

C.1 General-purpose optimiser

It is recommended to explore the working and the scope of a general-purpose optimiser by starting with an optimisation problem where the solution is known. The optimiser which is used for many examples in this book is `optim`; see, for example, Sections 3.4 and 4.5.

Here is an example to explore the working of `optim`. Consider the linear model $Y_i = \boldsymbol{\beta}^\top \mathbf{x}_i + \varepsilon_i$ with $\varepsilon_i \sim N(0, \sigma^2)$. Let R object y be the $N \times 1$ vector for the observed response, and let object `X` be the $N \times p$ design matrix.

Estimation of $\boldsymbol{\beta}$ and σ by least squares is standard:

```
beta <- solve( t(X)%*%X )%*%t(X)%*%y
s <- sqrt( (t(y-X%*%beta)%*%(y-X%*%beta))/(N-p) )
```

The corresponding maximum likelihood estimation can be implemented using `optim`. The default of this optimiser is minimisation. To maximise the log-likelihood we minimise $-1 \times$log-likelihood. For the case $p = 2$, this can be done as follows:

```
# Minus 1 times log-likelihood:
loglikelihood <- function(par){
  beta <- par[1:2]; sigma <- exp(par[3])
  return( -sum(log( dnorm(y,mean=X%*%beta,sd=sigma))) )
}
# Starting values:
par0 <- c(mean(y),0,log(sd(y)))
# Maximise:
max <- optim(par=par0, fn=loglikelihood, hessian=TRUE)
par <- max$par; par.var <- diag(solve(max$hessian))
# Delta method for sigma:
par.var[3] <- par.var[3]*exp(par[3])^2
# Results:
beta <- c(par[1:2],exp(par[3]))
s <- sqrt(par.var)
```

The function `dnorm` returns the density of a normal distribution for specified mean and standard deviation. The R object `max` includes the parameter values at which the maximum is reached, the value of the maximum, and the Hessian. Note the transformation to maximise over $\sigma \in (0, \infty)$, and the use of the delta method to estimate the standard error for estimated σ; see Section 3.7.2.

There are several options in `optim` with respect to the choice of algorithm, the maximum number of iterations, convergence criteria, etc.; see the R documentation. The default choice for the algorithm is the Nelder–Mead algorithm, which is a simplex method that works with values of the objective function only (as opposed to gradient-based methods).

Speed of the optimisation can be investigated by the function `system.time`. Here is an example to investigate the time needed for one function call:

```
system.time( loglikelihood(par0) )
```

C.2 Code for Chapter 2

The code for the Gompertz hazard model without left truncation is used for illustration. First, the hazard function `h` and the survivor function `S` are defined:

```
h <- function(t,lambda,xi){ lambda*exp(xi*t) }
S <- function(t,lambda,xi){
        exp(-lambda*xi^-1*(exp(xi*t)-1))
      }
```

Given the $N \times 1$ vectors `t` for event times and `delta` for right censoring, and the corresponding design matrix `X`, the coding of $-1\times$log-likelihood is given by

```
loglikelihood <- function(p){
 beta <- p[1:4];  xi <- p[5];  loglik <- rep(NA,N)
 for(i in 1:N){
  lambda.i <- exp(beta%*%X[i,1:4])
  loglik[i]<-
   log(S(t[i],lambda.i,xi)*h(t[i],lambda.i,xi)^delta[i])
 }
 return( -sum(loglik) )
}
```

This is the likelihood for a model with five parameters: intercept `beta[1]`, coefficients `beta[2:4]`, and the Gompertz parameter `xi`. The log-likelihood can be maximised using `optim`.

For all other models in Chapter 2 the code is written in a similar way. For the exponential model and the Weibull model fast routines are also available in the R package survival. Using this package the call for the Weibull model with five parameters is

```
m <- survreg(Surv(t,delta)~x1+x2+x3,
                data=dta,dist="weibull")
cat("Hazard parameters: ",c(-m$coeff/m$scale,1/m$scale))
```

The `survreg` function uses the accelerated failure time model representation of the Weibull hazard model. The second line of the code transforms the accelerated failure time parameters back to the parameters for the hazard model (2.6); see, for example, Collett (2003, Section 4.7).

The likelihood function in R for the models with the piecewise-constant hazards is essential the same as above, but is extended by additional looping over the time points in the grid.

The package survival includes the function coxph to fit the semi-parametric Cox hazard regression model for time-dependent data. This function requires so-called start-stop data in R; see, for example, Venables and Ripley (2002, Section 13.4).

C.3 Code for Chapter 3

Code is given for the three-state Weibull-baseline hazard model. The code for the other models can be derived using similar coding. The transition probability matrix $\mathbf{P}(t_1,t_2)$ is computed with the function

```
Pmatrix.Weib <- function(t1,t2,lambda,tau,nnodes,h){
 P <- diag(3)
 P[1,1] <- exp(-lambda[1]*(t2^tau[1]-t1^tau[1])-lambda[2]
                *(t2^tau[2]-t1^tau[2]))
 # Integrand for P[1,2]:
 integrand <- function(u){
    p11 <- exp(-lambda[1]*(u^tau[1]-t1^tau[1])-lambda[2]
               *(u^tau[2]-t1^tau[2]))
    q12 <- lambda[1]*tau[1]*u^(tau[1]-1)
    p22 <- exp(-lambda[3]*(t2^tau[3]-u^tau[3]))
    return(p11*q12*p22)
 }
 # Approximation using composite Simpson's rule:
 S <- integrand(t1)
 for(i in seq(1,nnodes,by=2)){
     x <- t1+h*i; S <- S+4*integrand(x)
 }
 for(i in seq(2,nnodes-1,by=2)){
     x <- t1+h*i; S <- S+2*integrand(x)
 }
 S <- S+integrand(t2)
 P[1,2] <- h*S/3
 # Rest of P-matrix:
 P[1,3] <- 1-P[1,1]-P[1,2]
 P[2,2] <- exp(-lambda[3]*(t2^tau[3]-t1^tau[3]))
 P[2,3] <- 1-P[2,2]
 return(P)
}
```

For maximum likelihood estimation, the coding of $-1\times$log-likelihood is

```
loglikelihood <- function(p){
 # Parameters:
 beta0 <- p[1:3]; tau <- exp(p[4:6])
 # Contribution per subject:
 loglik <-0
 for(i in 1:N){
  # Data subject i:
  data.i <- dta.split[[i]]
  O <- data.i$state; age <- data.i$age
  # Loop over individual follow up:
  for(j in 2:length(O)){
    # Time interval:
    t1 <- age[j-1]; t2 <- age[j]
    # Even number of nodes (with 4 as minimum):
    nnodes <- max(4,round((t2-t1)/h.int))
    nnodes <- ifelse(nnodes/2==round(nnodes/2),
                     nnodes,nnodes+1)
    h <- (t2-t1)/nnodes
    # P-matrix:
    P <- Pmatrix.Weib(t1,t2,exp(beta0),tau,nnodes,h)
    # Likelihood contribution:
    if(O[j]==D){
       q13 <- exp(beta0[2])*tau[2]*(t2-h)^(tau[2]-1)
       q23 <- exp(beta0[3])*tau[3]*(t2-h)^(tau[3]-1)
       contribution <- P[O[j-1],1]*q13+P[O[j-1],2]*q23
    }
    if(O[j]==censored){
       contribution <- P[O[j-1],1]+P[O[j-1],2]
    }
    if(O[j]==1|O[j]==2){
       contribution <- P[O[j-1],O[j]]
    }
    # Update likelihood:
    loglik <- loglik+log(contribution)
  }
 }
```

```
  return(-loglik)
}
```

The R object dta.split was created using the function split and allows quick access to individual-specific data (here index by i).

Object h.int is defined beforehand as the precision in the approximation of the integral. Objects D and censored are predefined and are used to code the dead state and right censoring, respectively. The log-likelihood is maximised using optim.

C.4 Code for Chapter 4

The following R code illustrates how the log-likelihood function can be implemented using the piecewise-constant approximation. The code is for a three-state progressive model with transition-specific Gompertz baseline hazards, but extension to more states is straightforward. The log-likelihood can be maximised using optim with default settings.

The coding of $-1\times$log-likelihood is

```
loglikelihood <- function(p){
  # Model parameters:
  beta <- p[1:3]; xi <- p[4:6]
  # Contribution per unit:
  loglik <- 0
  for(i in 1:N){
   # Data for subject i:
   data.i <- dta.split[[i]]
   O <- data.i$state; t <- data.i$age
   # Loop over observations for subject i:
   for(j in 2:length(O)){
     # Q and P matrix:
     Q <- matrix(0,3,3)
     t1 <- t[j-1]; t2 <- t[j]
     Q[1,2]<- exp(beta[1]+xi[1]*t1)
     Q[1,3]<- exp(beta[2]+xi[2]*t1); Q[1,1]<- -sum(Q[1,])
     Q[2,3]<- exp(beta[3]+xi[3]*t1); Q[2,2]<- -sum(Q[2,])
     P <- MatrixExp(mat=Q,t=t2-t1)
     # Likelihood contribution:
     death <- as.numeric(O[j]==D)
     loglik <- loglik+log((1-death)*P[O[j-1],O[j]] +
```

```
        (death)*(P[O[j-1],1]*Q[1,D]+P[O[j-1],2]*Q[2,D]))
    }
  }
  return(-loglik)
}
```

Note that age is the time scale t, and that for each observed interval (coded by t1 and t2 repeatedly) the generator matrix Q is kept constant. The function MatrixExp is used to compute transition probability matrices. This function is included in the msm package. Other options in this case are using a user-implemented closed-form solution, using eigenvalue decomposition with the R function eigen, or using the exmp package; see Appendix A.

C.5 Code for numerical integration

The following R code illustrates the trapezoidal rule and Gauss–Hermite quadrature. An illustration of the composite Simpson's rule can be found in the R code for Chapter 3; see above. Consider the integral given by

$$\mathcal{I} = \int_{-\infty}^{\infty} g(x) f_X(x|\mu,\sigma^2)dx,$$

where f_X is a probability density function for a normal distribution with mean μ and variance σ^2. For illustrative purposes, define $g(x) = x^2$, $\mu = 0$, and $\sigma^2 = 1$. For this specification, it follows that $\mathcal{I}$ is the expectation of a Chi-square distribution with degrees of freedom equal to 1, which means $\mathcal{I} = 1$.

The coding starts by specification of the functions g and f, and the integrand:

```
g <- function(x){ x^2 }
f <- function(x){ dnorm(x,mean=0,sd=1) }
integrand <- function(x){ g(x)*f(x) }
```

Code for the trapezoidal rule:

```
integrateTR <- function(nnodes,a,b){
   h <- (b-a)/nnodes
   grid <- seq(a,b,by=h); L <- length(grid)
   int <- integrand(grid)
   return( h/2*(int[1]+2*sum(int[2:(L-1)])+int[L]) )
}
```

With specified nnodes for the number of nodes, the integral is approximated by

```
integrateTR(nnodes=nnodes,a=-w,b=w)
```

where parameter w is used to define the support in the approximation symmetrically around $\mu = 0$.

For the Gauss–Hermite quadrature, the values of the nodes and the weights can be computed using the function gauss.quad in the R package statmod (Smyth, 2001). Code for the quadrature:

```
integrateGH <- function(nnodes,mu,sigma){
    # Nodes and weights for quadrature:
    quad <- gauss.quad(nnodes,"hermite")
    nodes <- quad$nodes; weights <- quad$weights
    # Compute integral:
    x <- sqrt(2)*sigma*nodes+mu
    return( (1/sqrt(pi))*weights%*%g(x) )
}
```

With the Guass–Hermite quadrature, the support is $(-\infty,\infty)$ by definition. In case of repeated function calls, computations can be speeded up by defining the nodes and the weights before applying the Gauss–Hermite quadrature repeatedly.

The comparison is more interesting for specifications of g for which Gaussian–Hermite does not give the exact solution. Compare, for instance, for various choices of the number of nodes the performance of the two methods using

```
# Function g:
g <- function(x){ x^2+log(abs(x)+1)  }
# Integration:
nnodes <- 15; w <- 5
integrateTR(nnodes=nnodes,a=-w,b=w)
integrateGH(nnodes=nnodes)
```

C.6 Code for Chapter 6

The following presents BUGS code for a progressive three-state survival model. Right censoring and known times of death are taken into account. The code is for an intercept-model only, but can be easily be extended to include covariate effects or frailties. As stated in Section 6.5, the code is an extension of the work in Welton and Ades (2005) and Pan et al. (2007); see also Van den Hout and Matthews (2009).

The data format is illustrated by

```
 id[] time[] current[] r[,1] r[,2] r[,3] r[,4]
  1    0.2     2         0     0     1     0
  2    0.9     1         0     0     1     0
  3    0.6     1         0     0     1     0
  4    1.0     2         0     1     0     0
  4    1.0     2         0     1     0     0
  4    0.5     2         0     0     1     0

  ...

  END
```

where `id[]` is the identifier, `time[]` is the length of the interval, `current[]` is the state at the start of the interval, and `r[,1] r[,2] r[,3] r[,4]` contains the coding for the state at the end of the interval; see Section 6.5.3 for the details. The BUGS code is given by

```
# Two sets of initial values for two chains:
list(beta=c(-2,-2,-2))
list(beta=c(-3,-1,-2))

# Model specification:
model{

# Approximating instant death by small time interval:
eps <- 1/12
# Maximum for entries Q to prevent numerical overflow:
maxQ <- 10

# Loop over, say, 100 records in the data:
for(i in 1:100) {

 # Transition intensities:
 G[i,1,2] <- min(maxQ,exp(beta[1]))
 G[i,1,3] <- min(maxQ,exp(beta[2]))
```

```
G[i,2,3] <- min(maxQ,exp(beta[3]))

# Transition probabilities for observed interval:
P[i,1,1] <- exp(-(G[i,1,2]+G[i,1,3])*time[i])
P[i,1,2] <- (G[i,1,2]/(G[i,1,2]+G[i,1,3]-G[i,2,3]))*
    exp(-G[i,2,3]*time[i])*
    (1-exp(-(G[i,1,2]+G[i,1,3]-G[i,2,3])*time[i]))
P[i,1,3] <- 1-P[i,1,1]-P[i,1,2];   P[i,2,1]<-0
P[i,2,2] <- exp(-G[i,2,3]*time[i])
P[i,2,3] <- 1-P[i,2,2]

# Transition probabilities for length eps:
Peps[i,1,1] <- exp(-(G[i,1,2]+G[i,1,3])*eps)
Peps[i,1,2] <- (G[i,1,2]/(G[i,1,2]+G[i,1,3]-
     G[i,2,3]))*exp(-G[i,2,3]*eps)*
     (1-exp(-(G[i,1,2]+G[i,1,3]-G[i,2,3])*eps))
Peps[i,1,3] <- 1-Peps[i,1,1]-Peps[i,1,2]
Peps[i,2,1] <- 0 ;  Peps[i,2,2 ]<- exp(-G[i,2,3]*eps)
Peps[i,2,3] <- 1-Peps[i,2,2]

# Transition probabilities for length interval-eps:
Pteps[i,1,1] <- exp(-(G[i,1,2]+G[i,1,3])*(time[i]-eps))
Pteps[i,1,2] <- (G[i,1,2]/(G[i,1,2]+G[i,1,3]-
   G[i,2,3]))*exp(-G[i,2,3]*(time[i]-eps))*
   (1-exp(-(G[i,1,2]+G[i,1,3]-G[i,2,3])*(time[i]-eps)))
Pteps[i,1,3] <- 1-Pteps[i,1,1]-Pteps[i,1,2]
Pteps[i,2,1] <- 0
Pteps[i,2,2] <- exp(-G[i,2,3]*(time[i]-eps))
Pteps[i,2,3] <- 1-Pteps[i,2,2]

# State at end of interval:
dead[i] <- equals(r[i,3], 1); alive[i] <- 1-dead[i]
cens[i] <- equals(r[i,4],1);  notcens[i] <- 1-cens[i]

# Define the transition matrix taking death
# and right censoring into account:
pd1[i]<-Pteps[i,1,1]*Peps[i,1,3] +
    Pteps[i,1,2]*Peps[i,2,3]
PP[i,1,1]<-alive[i]*notcens[i]*P[i,1,1] +
    dead[i]*P[i,1,1]*(1-pd1[i])/(1-P[i,1,3])
```

```
  PP[i,1,2]<-alive[i]*notcens[i]*P[i,1,2] +
      dead[i]*P[i,1,2]*(1-pd1[i])/(1-P[i,1,3])
  PP[i,1,3]<-alive[i]*P[i,1,3] + dead[i]*pd1[i]
  PP[i,1,4]<-cens[i]*(1-P[i,1,3])

  pd2[i] <- Pteps[i,2,2]*Peps[i,2,3]
  PP[i,2,1] <- 0
  PP[i,2,2] <- alive[i]*notcens[i]*P[i,2,2] +
       dead[i]*P[i,2,2]*(1-pd2[i])/(1-P[i,2,3])
  PP[i,2,3] <- alive[i]*P[i,2,3] + dead[i]*pd2[i]
  PP[i,2,4] <- cens[i]*(1-P[i,2,3])

  # Multinomial likelihood for data:
  r[i,1:4] ~ dmulti(PP[i,current[i],1:4],1)
 }

 # Priors for intensities parameters:
 # Uniform(-15,10) can be considered vague
 # in most settings (else adjust):
 for(j in 1:3){ beta[j] ~ dunif(-15,10) }
 }
```

Bibliography

Aalen, O. O., Borgan, O., and Gjessing, H. (2008). *Survival and Event History Analysis*. New York: Springer.

Aalen, O. O., Farewell, V. T., De Angelis, D., Day, N. E., and Gill, O. N. (1997). A Markov model for HIV disease progression including the effect of HIV diagnosis and treatment: application to AIDS prediction in England and Wales. *Statistics in Medicine*, 16:2191–2210.

Agresti, A. (2002). *Categorical Data Analysis (2nd edition)*. Hoboken: Wiley.

Aguirre-Hernandez, R. and Farewell, V. T. (2002). A Pearson-type goodness-of-fit test for stationary and time-continuous Markov regression models. *Statistics in Medicine*, 21:1899–1911.

Aitkin, M. (1999). A general maximum likelihood analysis of variance components in generalized linear models. *Biometrics*, 55:117–128.

Aitkin, M., Darnell, R., Francis, B., and Hinde, J. (2009). *Statistical Modelling in R*. Oxford: Clarendon Press.

Akaike, H. (1974). A new look at the statistical model identification. *IEEE Transactions on Automatic Control*, 19:716–723.

Albarran, I., Ayuso, M., Guillén, M., and Monteverde, M. (2005). A multiple state model for disability using the decomposition of death probabilities and cross-sectional data. *Communications in Statistics —Theory and Methods*, 34:2063–2075.

Alioum, A. and Commenges, D. (2001). MKVPCI: a computer program for Markov models with piecewise constant intensities and covariates. *Computer Methods and Programs in Biomedicine*, 64:109–119.

Andersen, P. K. (2002). Multi-state models. *Statistical Methods in Medical Research*, 11:89–90.

Andersen, P. K. (2013). Decomposition of number of life years lost according to causes of death. *Statistics in Medicine*, 32:5278–5285.

Andersen, P. K., Borgan, O., Gill, R. D., and Keiding, N. (1993). *Statistical Models Based on Counting Processes*. New York: Springer.

Anderson, E., Bai, Z., Bischof, C., Blackford, S., Demmel, J., Dongarra, J., Du Croz, J., Greenbaum, A., Hammarling, S., McKenney, A., and Sorensen, D. (1999). *LAPACK Users' Guide (3rd edition)*. Philadelphia: SIAM.

Bacchetti, P., Boylan, R. D., Terrault, N. A., Monto, A., and Berenguer, M. (2010). Non-Markov multistate modeling using time-varying covariates, with application to progression of liver fibrosis due to hepatitis C following liver transplant. *The International Journal of Biostatistics*, 6.

Bartolucci, F., Farcomeni, A., and Pennoni, F. (2012). *Latent Markov Models for Longitudinal Data*. London: Chapman & Hall/CRC.

Bender, F., Augustin, T., and Blettner, M. (2005). Generating survival times to simulate Cox proportional hazards models. *Statistics in Medicine*, 24:1713–1723.

Bernardo, J. M. and Smith, A. (2000). *Bayesian Theory*. Chichester: Wiley.

Beyersmann, J., Schumacher, M., and Allignol, A. (2012). *Competing Risks and Multistate Models with R*. New York: Springer.

Brayne, C. and Calloway, P. (1990). The case identification of dementia in the community: a comparison of methods. *International Journal of Geriatric Psychiatry*, 5:309–316.

Brayne, C., McCracken, C., and Matthews, F. E. (2006). Cohort profile: the Medical Research Council Cognitive Function and Ageing Study (CFAS). *International Journal of Epidemiology*, 35:1140–1145.

Broyden, C. G. (1970). The convergence of a class of double-rank minimization algorithms. *Journal of the Institute of Mathematics and Its Applications*, 6:76–90.

Bull, K. and Spiegelhalter, D. J. (1997). Survival analysis in observational studies. *Statistics in Medicine*, 16:1041–1074.

Bureau, A., Shiboski, S., and Hughes, J. P. (2003). Applications of continuous time hidden Markov models to the study of misclassified disease outcomes. *Statistics in Medicine*, 22:441–462.

Buter, T. C., Van den Hout, A., Matthews, F. E., Larsen, J. P., Brayne, C., and Aarsland, D. (2008). Dementia and survival in Parkinson's disease—a twelve year population study. *Neurology*, 70:1017–1022.

Cai, L. and Lubitz, J. (2007). Was there compression of disability for older Americans from 1992 to 2003? *Demography*, 44:479–495.

Cai, L., Lubitz, J., Hayward, M. D., Hagedorn, A., Saito, Y., and Crimmins, E. (2010). Estimation of multi-state life table functions and their variability from complex survey data using the SPACE Program. *Demographic Research*, 22:129–158.

Carlin, B. P. and Louis, T. A. (2009). *Bayesian Methods for Data Analysis. Third Edition*. Boca Raton, FL: CRC Press.

Casella, G. and Berger, R. L. (2002). *Statistical Inference (2nd edition)*. Australia: Duxbury.

Cole, B. F., Bonetti, M., Zaslavsky, A. M., and Gelber, R. D. (2005). A multistate Markov chain model for longitudinal, categorical quality-of-life data subject to non-ignorable missingness. *Statistics in Medicine*, 24:2317–2334.

Collett, D. (2003). *Modelling Survival Data in Medical Research (2nd edition)*. London: Chapman & Hall/CRC.

Commenges, D. (1999). Multi-state models in epidemiology. *Lifetime Data Analysis*, 5:315–317.

Commenges, D. (2002). Inference for multi-state models from interval-censored data. *Statistical Methods in Medical Research*, 11:167–182.

Commenges, D., Letenneur, L., Joly, P., Alioum, A., and Dartigues, J.-F. (1998). Modelling age-specific risk: application to dementia. *Statistics in Medicine*, 17:1973–1988.

Cook, R. J. and Lawless, J. F. (2014). Statistical issues in modeling chronic disease in cohort studies. *Statistics in Biosciences*, 6:127–161.

Cox, D. R. (1972). Regression models and life tables (with discussion). *Journal of the Royal Statistical Society: Series B (Statistical Methodology)*, 74:187–220.

Cox, D. R. and Miller, H. D. (1965). *The Theory of Stochastic Processes*. London: Chapman & Hall.

Cox, D. R. and Oakes, D. (1984). *Analysis of Survival Data*. London: Chapman & Hall/CRC.

Crowder, M. J. (2012). *Multivariate Survival Analysis and Competing Risks*. London: Chapman & Hall/CRC.

Davis, B. A., Heathcote, C. R., and O'neill, T. J. (2002). Estimating and interpolating a Markov chain from aggregate data. *Biometrika*, 89:95–110.

Davison, A. C. and Ramesh, N. I. (1996). Some models for discretized series of events. *Journal of the American Statistical Association*, 91:601–609.

De Lau, L. M., Schipper, C. M., Hofman, A., Koudstaal, P. J., and Breteler, M. M. B. (2005). Prognosis of Parkinson disease: risk of dementia and mortality: the Rotterdam Study. *Archives of Neurology*, 62:1265–1269.

Dempster, A. P., Laird, N. M., and Rubin, D. B. (1977). Maximum likelihood from incomplete data via the EM algorithm. *Journal of the Royal Statistical Society: Series B (Statistical Methodology)*, 39:1–38.

Diggle, P. J., Heagerty, P., Liang, K.-Y., and Zeger, S. (2002). *Analysis of Longitudinal Data, Second Edition.* Oxford: Oxford University Press.

Dinse, G. E. (1988). Estimating tumor incidence rates in animal carcinogenicity experiments. *Biometrics*, 44:405–415.

Duchateau, L. and Janssen, P. (2008). *The Frailty Model.* New York: Springer.

Eilers, P. H. C. and Marx, B. D. (1996). Flexible smoothing with B-splines and penalties. *Statistical Science*, 11:89–121.

Einbeck, J. and Hinde, J. (2009). *Nonparametric maximum likelihood estimation for random effect models in R. Vignette to R package npmlreg version 0.44.*

Faddy, M. J. (1976). A note on the general time-dependent stochastic compartmental model. *American Journal of Epidemiology*, 32:443–448.

Ferrucci, L., Izmirlian, G., Leveille, S., Phillips, C. L., Corti, M. C., Brock, D. B., and Guralnik, J. M. (1999). Smoking, physical activity, and active life expectancy. *American Journal of Epidemiology*, 149:645–653.

Fletcher, R. (1970). A new approach to variable metric methods. *Computer Journal*, 13:317–322.

Folstein, M. F., Folstein, S. E., and McHugh, P. R. (1975). Mini-mental state: a practical method for grading the state of patients for the clinician. *Journal of Psychiatric Research*, 12:189–198.

Foucher, Y., Giral, M., Soulillou, J., and Daures, J. (2010). A flexible semi-Markov model for interval-censored data and goodness-of-fit testing. *Statistical Methods in Medical Research*, 12:127–145.

Fox, J.-P. (2010). *Bayesian item response modeling*. New York: Springer.

Frydman, H. (1984). Maximum likelihood estimation in the mover-stayer model. *Journal of the American Statistical Association*, 79:632–638.

Frydman, H. and Szarek, M. (2009). Nonparametric estimation in a Markov illness-death process from interval censored observations with missing intermediate transition status. *Biometrics*, 65:143–151.

Gelman, A. (2006). Prior distributions for variance parameters in hierarchical models. *Bayesian Analysis*, 1:515–533.

Gelman, A., Carlin, J. B., Stern, H. S., and Rubin, D. B. (2004). *Bayesian Data Analysis*. London: Chapman & Hall.

Gelman, A. and Rubin, D. B. (1992). Inference from iterative simulation using multiple sequences. *Statistical Science*, 7:457–511.

Geman, S. and Geman, D. (1984). Stochastic relaxation, Gibbs distributions, and the Bayesian restoration of images. *IEEE Trans. Pattern Analysis and Machine Intelligence*, 6:721–741.

Gentleman, R. C., Lawless, J. F., Lindsey, J. C., and Yan, P. (1994). Multistate Markov models for analysing incomplete disease data with illustrations for HIV disease. *Statistics in Medicine*, 13:805–821.

Gillaizeau, F., Dantan, E., Giral, M., and Foucher, F. (2015). A multistate additive relative survival semi-Markov model. *Statistical Methods in Medical Research*, Advance online publication. doi: 10.1177/0962280215586456.

Glaeskens, G. and Hjort, N. (2008). *Model Selection and Model Averaging*. Cambridge: Cambridge University Press.

Goldfarb, D. (1970). A family of variable metric methods derived by variational means. *Mathematics of Computation*, 24:23–26.

Goulet, V., Dutang, C., Maechler, M., Firth, D., Shapira, M., and Stadelmann, M. (2014). *expm: matrix exponential. R package version 0.99-1.1.*

Grag, M. L., Rao, R., and Redmond, C. K. (1970). Maximum-likelihood estimation of the parameters of the Gompertz survival function. *Journal of the Royal Statistical Society: Series C (Applied Statistics)*, 19:152–159.

Grüger, J., Kay, R., and Schumacher, M. (1991). The validity of inferences based on incomplete observations in disease state models. *Biometrics*, 47:595–605.

Hajducek, D. M. and Lawless, J. F. (2012). Duration analysis in longitudinal studies with intermittent observation times and losses to followup. *The Canadian Journal of Statistics*, 40:1–21.

Hawkins, D. L. and Han, C.-P. (2000). Estimating transition probabilities from aggregate samples plus partial transition data. *Biometrics*, 56:848–854.

Hougaard, P. (2000). *Analysis of Multivariate Survival Data*. New York: Springer.

Hubbard, R., Inoue, L. Y. T., and Fann, J. R. (2008). Modeling nonhomogeneous Markov processes via time transformation. *Biometrics*, 64:843–850.

Imai, K. and Soneji, S. (2007). On the estimation of disability-free life expectancy: Sullivans method and its extension. *Journal of the American Statistical Association*, 102:1199–1211.

Izmirlian, G., Brock, D., Ferrucci, L., and Phillips, C. (2000). Active life expectancy from annual follow-up data with missing responses. *Biometrics*, 56:244–248.

Jackson, C. (2011). Multi-state models for panel data: the msm package for R. *Journal of Statistical Software*, 38.

Jackson, C. H. and Sharples, L. D. (2002). Hidden Markov models for the onset and progression of bronchiolitis obliterans syndrome in lung transplant recipients. *Statistics in Medicine*, 21:113–128.

Jackson, C. H., Sharples, L. D., Thompson, S. G., Duffy, S. W., and Couto, E. (2003). Multi-state Markov models for disease progression with classification error. *Statistician*, 52:193–209.

Jagger, C., Cox, B., Le Roy, S., and the EHEMU team (2007). *Health Expectancy Calculation by the Sullivan Method. Third Edition. European Health Expectancy Monitoring Unit (EHEMU) Technical Report.*

Jennrich, R. I. and Bright, P. B. (1976). Fitting systems of linear differential equations using computer generated exact derivatives. *Technometrics*, 18:385–392.

Johnson, N. L., Balakrishnan, N., and Kotz, S. (1994). *Continuous Univariate Distributions: Volume 1.* New York: Wiley.

Johnson, V. E. and Albert, J. H. (1999). *Ordinal Data Modeling.* New York: Springer.

Joly, P., Commenges, D., and Letenneur, L. (1998). A penalized likelihood approach for arbitrarily censored and truncated data: application to age-specific incidence of dementia. *Biometrics*, 54:185–194.

Joly, P., Durand, C., Helmer, C., and Commenges, D. (2009). Estimating life expectancy of demented and institutionalized subjects from interval-censored observations of a multi-state model. *Statistical Modelling*, 9:345–360.

Kalbfleisch, J. and Lawless, J. F. (1985). The analysis of panel data under a Markov assumption. *Journal of the American Statistical Association*, 80:863–871.

Kalbfleisch, J. and Lawless, J. F. (1988). Likelihood analysis of multi-state models for disease incidence and mortality. *Statistics in Medicine*, 7:149–160.

Kalbfleisch, J. D. and Prentice, R. L. (2002). *The Statistical Analysis of Failure Time Data (2nd edition).* New York: Wiley.

Kapetanakis, V., Matthews, F. E., and Van den Hout, A. (2013). A semi-

Markov model for stroke with piecewise-constant hazards in the presence of left-, right-, and interval-censoring. *Statistics in Medicine*, 32:697–713.

Kay, R. (1986). A Markov model for analysing cancer markers and disease states in survival studies. *Biometrics*, 42:855–865.

Kneib, T. and Hennerfeind, A. (2008). Bayesian semi parametric multi-state models. *Statistical Modelling*, 8:169–198.

Korn, E. L., Graubard, B. I., and Midhune, D. (1997). Time-to-event analysis of longitudinal follow-up of a survey: choice of time scale. *American Journal of Epidemiology*, 145:72–80.

Kulkarni, V. G. (2011). *Introduction to Modeling and Analysis of Stochastic Systems (2nd edition)*. New York: Springer.

Lagakos, S. W., Sommer, C. J., and Zelen, M. (1978). Semi-Markov models for partially censored data. *Biometrika*, 65:311–317.

Lenart, A. (2014). The moments of the Gompertz distribution and maximum likelihood estimation of its parameters. *Scandinavian Actuarial Journal*, 2014:255–277.

Lièvre, A., Brouard, N., and Heathcote, C. (2003). The estimation of health expectancies from cross-longitudinal surveys. *Mathematical Population Studies*, 10:211–248.

Lindsay, B. G. (1995). *Mixture Models: Theory, Geometry and Applications*. NSF-CBMS Regional Conference Series in Probability and Statistics, Vol. 5. Hayward, CA: Institute of Mathematical Statistics.

Lindsey, J. C. and Ryan, L. (1994). A comparison of continuous- and discrete-time three-state models for rodent tumorigenicity experiments. *Environmetal Health Perspectives Supplements*, 102:9–17.

Little, R. J. A. and Rubin, D. B. (2002). *Statistical Analysis with Missing Data (2nd edition)*. New York: Wiley.

Lunn, D., Jackson, C., Best, N., Thomas, A., and Spiegelhalter, D. (2013). *The BUGS Book*. Boca Raton, FL: CRC Press.

Lunn, D., Spiegelhalter, D., Thomas, T., and Best, N. (2009). The BUGS project: evolution, critique and future directions. *Statistics in Medicine*, 28:3049–3067.

Mandel, M. (2013). Simulation-based confidence intervals for functions with complicated derivatives. *The American Statistician*, 67:76–81.

Marshall, G., Guo, W., and Jones, R. H. (1995). MARKOV: a computer program for multi-state Markov models with covariables. *Computer Methods and Programs in Biomedicine*, 47:147–156.

Marshall, G. and Jones, R. H. (1995). Multi-state models and diabetic retinopathy. *Statistics in Medicine*, 14:1975–1995.

Metropolis, N., Rosenbluth, A. W., Rosenbluth, M. N., Teller, A. H., and Teller, E. (1953). Equation of state calculation by fast computing machines. *Journal of Chemical Physics*, 21:1087–1092.

Molenberghs, G. and Verbeke, G. (2005). *Models for Discrete Longitudinal Data.* New York: Springer.

Moler, C. and Van Loan, C. (2003). Nineteen dubious ways to compute the exponential of a matrix, twenty-five years later. *SIAM Review*, 45:3–49.

Muthén, B. and Asparouhov, T. (2009). Growth mixture modeling: analysis with non-Gaussian random effects. In Fitzmaurice, G., Davidian, M., Verbeke, G., and Molenberghs, G., editors, *Longitudinal Data Analysis*, pp. 143–166. Boca Raton, FL: Chapman & Hall/CRC Press.

Nelder, J. A. and Mead, R. (1965). A simplex algorithm for function minimization. *Computer Journal*, 7:308–313.

Newman, S. A. (1988). A Markov process interpretation of Sullivan's index of morbidity and mortality. *Statistics in Medicine*, 7:787–794.

Norris, J. R. (1997). *Markov Chains.* Cambridge: Cambridge University Press.

O'Hagan, A., Stevenson, M., and Madan, J. (2007). Monte Carlo probabilistic sensitivity analysis for patient level simulation models: efficient estimation of mean and variance using ANOVA. *Health Economics*, 16:1009–1023.

Omar, R. Z., Stallard, N., and Whitehead, N. (1995). A parametric multistate model for the analysis of carcinogenicity experiments. *Lifetime Data Analysis*, 1:327–346.

Pan, S. L., Wu, H. M., Yen, A. M. F., and Chen, T. (2007). A Markov regression random-effects model for remission of functional disability in patients following a first stroke: a Bayesian approach. *Statistics in Medicine*, 26:5335–5353.

Pelzer, B., Eisinga, R., and Franses, P. H. (2001). Estimating transition probabilities from a time series of independent cross sections. *Statistica Neerlandica*, 55:249–262.

Plummer, M. (2003). JAGS: a program for analysis of Bayesian graphical models using Gibbs sampling. Vienna, Austria. ISSN 1609-395X. In *Proceedings of the 3rd International Workshop on Distributed Statistical Computing (DSC 2003).*

Plummer, M. (2008). Penalized loss functions for Bayesian model comparison. *Biostatistics*, 9:523–539.

Plummer, M., Best, N., Cowles, K., and Vines, K. (2006). CODA: convergence diagnosis and output analysis for MCMC. *R News*, 6:7–11.

Pollard, J. H. and Valkovics, E. J. (1992). The Gompertz distribution and its applications. *Genus*, 48:15–28.

Putter, H., Fiocco, M., and Geskus, R. B. (2007). Tutorial in biostatistics: competing risks and multi-state models. *Statistics in Medicine*, 26:2389–2430.

Putter, H. and Van Houwelingen, H. C. (2015). Frailties in multi-state models: Are they identifiable? Do we need them? *Statistical Methods in Medical Research*, 24:675–692.

R Core Team (2013). *R: A Language and Environment for Statistical Computing*. Vienna, Austria: R Foundation for Statistical Computing.

Rice, J. A. (1995). *Mathematical Statistics and Data Analysis*. Belmont: Duxbury Press.

Rondeau, V., Filleul, L., and Joly, P. (2006). Nested frailty models using maximum penalized likelihood estimation. *Statistics in Medicine*, 25:4036–4052.

Ross, S. M. (2010). *Introduction to Probability Models (10th edition)*. Amsterdam: Academic Press.

Satten, G. A. (1999). Estimating the extent of tracking in interval-censored chain-of-events data. *Biometrics*, 55:1228–1231.

Satten, G. A. and Longini, I. M. (1996). Markov chains with measurement error: estimating the 'true' course of a marker of the progression of human immunodeficiency virus disease (with discussion). *Applied Statistics*, 45:275–309.

Schwartz, G. (1978). Estimating the dimension of a model. *The Annals of Statistics*, 6:461–464.

Shanno, D. F. (1970). Conditioning of quasi-Newton methods for function minimization. *Mathematics of Computation*, 24:647–656.

Sharples, L. D. (1993). Use of the Gibbs sampler to estimate transition rates between grades of coronary disease following cardiac transplantation. *Statistics in Medicine*, 12:1155–1169.

Sharples, L. D., Jackson, C. H., Parameshwar, J., Wallwork, J., and Large, S. R. (2003). Diagnostic accuracy of coronary angiography and risk factors for post-heart-transplant cardiac allograft vasculopathy. *Transplantation*, 76:679–682.

Silverman, B. W. (1986). *Density Estimation*. London: Chapman & Hall.

Smyth, G. (2001). *Statmod. R package version version 1.4.18.*

Spiegelhalter, D. J. and Best, N. G. (2003). Bayesian approaches to multiple sources of evidence and uncertainty in complex cost-effectiveness modelling. *Statistics in Medicine*, 22:3687–3709.

Spiegelhalter, D. J., Best, N. G., Carlin, B., and Van der Linde, A. (2002). Bayesian measures of model complexity and fit (with discussion). *Journal of the Royal Statistical Society: Series B (Statistical Methodology)*, 64:583–640.

Steptoe, A., Breeze, E., Banks, J., and Nazroo, J. (2013). Cohort Profile: The English Longitudinal Study of Ageing. *International Journal of Epidemiology*, 42:1640–1648.

Sullivan, D. (1971). A single index of mortality and morbidity. *HSMHA Health Reports*, 86:347–354.

Sun, J. (2006). *The Statistical Analysis of Interval-Censored Failure Time Data*. New York: Springer.

Sutradhar, R. and Cook, R. (2008). Analysis of interval-censored data from clustered multistate processes: application to joint damage in psoriatic arthritis. *Journal of the Royal Statistical Society: Series C (Applied Statistics)*, 57:553–566.

Tanner, M. A. (1996). *Tools for Statistical Inference: Methods for the Exploration of Posterior Distributions and Likelihood Functions*. New York: Springer.

Terneau, T. M. and Grambsch, P. M. (2000). *Modeling Survival Data. Extending the Cox Model*. New York: Springer.

Titman, A. C. (2009). Computation of the asymptotic null distribution of goodness-of-fit tests for multi-state models. *Lifetime Data Analysis*, 15:519–533.

Titman, A. C. (2011). Flexible nonhomogeneous Markov models for panel observed data. *Biometrics*, 67:780–787.

Titman, A. C. and Sharples, L. D. (2010a). Model diagnostics for multi-state models. *Statistical Methods in Medical Research*, 19:621–651.

Titman, A. C. and Sharples, L. D. (2010b). Semi-Markov models with phase-type sojourn distributions. *Biometrics*, 66:742–752.

Tom, B. D. M. and Farewell, V. T. (2011). Intermittent observation of time-dependent explanatory variables: a multistate modelling approach. *Statistics in Medicine*, 30:3520–3531.

Van de Kassteele, J., Hoogenveen, R., Engelfriet, P., Baal, P., and Boshuizen,

H. (2012). Estimating net transition probabilities from cross-sectional data with application to risk factors in chronic disease modeling. *Statistics in Medicine*, 31:533–543.

Van den Hout, A., Fox, J.-P., and Klein Entink, R. (2015). Bayesian inference for an illness-death model for stroke with cognition as a latent time-dependent risk factor. *Statistical Methods in Medical Research*, 24:769–787.

Van den Hout, A., Jagger, C., and Matthews, F. E. (2009). Estimating life expectancy in health and ill health by using a hidden Markov model. *Journal of the Royal Statistical Society: Series C (Applied Statistics)*, 58:449–465.

Van den Hout, A. and Matthews, F. E. (2008). Multi-state analysis of cognitive ability data: a piecewise-constant model and a Weibull model. *Statistics in Medicine*, 27:5440–5455.

Van den Hout, A. and Matthews, F. E. (2009). Estimating dementia-free life expectancy for Parkinson's patients using Bayesian inference and microsimulation. *Biostatistics*, 10:729–743.

Van den Hout, A. and Matthews, F. E. (2010). Estimating stroke-free and total life expectancy in the presence of non-ignorable missing values. *Journal of the Royal Statistical Society: Series A (Statistics in Society)*, 173:331–349.

Van den Hout, A., Muniz-Terrera, G., and Matthews, F. E. (2013). Change point models for cognitive tests using semi-parametric likelihood. *Computational Statistics and Data Analysis*, 57:684–698.

Van den Hout, A., Ogurtsova, E., Gampe, J., and Matthews, F. E. (2014). Investigating healthy life expectancy using a multi-state model in the presence of missing data and misclassification. *Demographic Research*, 30:1219–1244.

Van den Hout, A. and Tom, B. (2013). Survival analysis and the frailty model. In Scott, M., Simonoff, J., and Marx, B., editors, *The SAGE Handbook of Multilevel Modeling*, pp. 541–558. Thousand Oaks, CA: Sage Publications.

Venables, W. N. and Ripley, B. D. (2002). *Modern Applied Statistics with S (4th edition)*. New York: Spinger.

Vermunt, J. K., Langeheine, R., and Böckenholt, U. (1999). Discrete-time discrete-state latent Markov models with time-constant and time-varying covariates. *Journal of Educational and Behavioral Statistics*, 24:179–207.

Vermunt, J. K. and Magidson, J. (2002). Latent class cluster analysis. In Hagenaars, J. and McCutcheon, A., editors, *Applied Latent Class Analysis*, pp. 89–106. Cambridge: Cambridge University Press.

Wang, Y. (2010). Maximum likelihood computation for fitting semiparametric mixture models. *Statistics and Computing*, 20:75–86.

Wei, L. J. (1992). The accelerated failure time model: a useful alternative to the Cox regression model in survival analysis. *Statistics in Medicine*, 11:1871–1879.

Weiss, G. H. and Zelen, M. (1965). A semi-Markov model for clinical trials. *Journal of Applied Probability*, 2:269–285.

Welton, N. and Ades, A. D. (2005). Estimation of Markov chain transition probabilities and rates from fully and partially observed data: uncertainty propagation, evidence synthesis, and model calibration. *Medical Decision Making*, 25:633–645.

Willekens, F. (2005). Biographic forecasting: bridging the mirco-macro gap in population forecasting. *New Zealand Population Review*, 3:77–124.

Willekens, F. (2014). *Multistate Analysis of Life Histories with R*. New York: Springer.

Wu, L. (2010). *Mixed Effects Models for Complex Data*. Boca Raton, FL: Chapman & Hall/CRC.

Zhang, M. and Schaubel, D. E. (2011). Estimating differences in restricted mean lifetime using observational data subject to dependent censoring. *Biometrics*, 67:740–749.

Zinn, S., Himmelspach, J., Gampe, J., and Uhrmacher, A. M. (2009). MIC-CORE: a tool for microsimulation. In *Proceedings of the 2009 Winter Simulation Conference*, pp. 992–1002. Piscataway, NJ: IEEE.

Index